TETRAPOD ZOOLOGY

bookone

DARREN NAISH

Edited and Typeset by Jonathan Downes,
Cover by C. M. Koseman
Indexed by Daniel Taylor, Graham Inglis, Oll Lewis
Layout by SPiderKaT for CFZ Communications
Using Microsoft Word 2000, Microsoft , Publisher 2000, Adobe Photoshop CS.

First published in Great Britain by CFZ Press

CFZ Press
Myrtle Cottage
Woolsery
Bideford
North Devon
EX39 5QR

Second edition 2010

ISBN: 978-1-905723-61-4

Tetrapod Zoology
est. 2006
scienceblogs.com/tetrapodzoology

Contents

Foreword

When Darren asked me if I'd mind writing the foreword to this book, I was fearing something of a slog through esoteric and dry script, but was pleasantly surprised! This epic tome represents the best of a whole year's worth of blogging, and a wealth of intriguing information, surprising science, tantalising possibilities and occasional freaky fantasy.

There are many themes here certain to capture the attention; giant bear-eating reticulated pythons, man-eating birds of prey, alien big cats and new discoveries about dinosaurs. The book though caught my attention with its honesty and enthusiasm, and is clearly written by a man who is endlessly fascinated by the natural world, and his own ceaseless desire to find out everything he can about it. In that, he is a man after my own heart!

Steve Backshall

Steve Backshall is a British naturalist, writer and television presenter, notably on Children's BBC's *The Really Wild Show*, *Deadly 60*, BBC TV's *Lost Land of the Volcano*, *Lost Land of the Jaguar*, *Expedition Alaska*, *Wilderness St Kilda* and *Expedition Borneo*, and on National Geographic Channel's *EarthPulse* series. He is also an author with the Rough Guides. In 2009, Backshall was nominated for three BAFTAs, Best Children's Television Presenter, Best Children's series, and Best Factual series for *Lost Land of the Jaguar*, as well as an Emmy for *Expedition Borneo*.

Introduction

Blogging has revolutionised the way we communicate our interests and spread news. This book is a compilation of various articles from my blog, Tetrapod Zoology (currently hosted at www.scienceblogs.com/tetrapodzoology). As of early 2010, Tet Zoo – as it's affectionately known – is in its fifth year. It's become reasonably popular (it has a daily readership of several thousand) and is now well known internationally. Or, it is, at least, among people interested in zoology and in scientific blogging. Since January 2007 Tet Zoo has been hosted on ScienceBlogs, a community of science-based blogs. I've made many new friends through Tet Zoo, have become involved in various exciting academic projects and partnerships, and have even landed a few paying jobs. So, welcome to the world of Tet Zoo: amphibians, reptiles, birds and mammals (the tetrapods), living and fossil. Their evolution, ecology, behaviour and biology. Think killer eagles, dinosaurs, giant caimans, mystery cats and lake monsters.

Fig. 1. The author on a windy beach.

I'm a vertebrate palaeontologist: I study dinosaurs, pterosaurs and other fossil reptiles that lived during the Mesozoic Era. But I'm also very much interested in modern animals, and also in the fossil animals of the Palaeozoic and Cenozoic (the two geological eras that, respectively, came before and after the Mesozoic). For as long as I've been able to, I've been conducting academic research on these creatures, and writing about them: blogging has given me the ultimate freedom to communicate my thoughts and research to the masses. I do blog about my own research – about the dinosaur and pterosaur stuff – but I also blog about other animals when I find them interesting, or when I think I have something interesting to say about them.

Tet Zoo started its life on Saturday 21st January 2006 when, entirely on a whim, I decided to start my own blog. I'd already decided what I was going to do: to write articles on the animals that interested me – the tetrapods – and to discuss their diversity, ecology, evolution and behaviour. January 2006 was just about the worse time to do all this, given that I was scheduled to complete my PhD in April of that year. And I was already too busy with additional commitments. But I did it anyway: the very first thing I posted (other than a portrait photo) was one of my zoology-themed Christmas cards (Fig. 2). While lying in bed later that night I felt the over-riding urge to get up, switch the computer on and write the first of my patented over-long, tediously detailed blog essays. It was on giant killer eagles, a favourite subject of mine, and one that I was to return to several times later on in the year. It marked the beginning of a new and highly rewarding phase in my life.

The articles compiled in this book are among the best from what I term Tet Zoo ver 1: the pre-ScienceBlogs version of Tet Zoo. Dating entirely from 2006, the articles thus range from slightly to highly dated (yikes, where did those four years go?). New work and new discoveries mean that some of the assertions made in these 2006 articles have been superseded. Needless to say, taking account of this new work would have meant a thorough re-write of the original articles. Rather than do this, I've mostly been lazy and have left them as they were. However, in some places I've added small updates that help show the reader what has happened since. There are several places where small updates just wouldn't do the job though, so – in the interests of self-correction – please keep the following in mind.

Updates and postscripts

Chapter 1 discusses the proposal that a large eagle might have killed the australopithecine specimen known as the Taung child, and refers to a technical study that is soon to appear in the literature. That study is Berger (2006), and note also Berger & McGraw's (2003) article. Giraffe taxonomy – discussed in Chapter 2 – was overhauled in 2007 (Brown *et al.* 2007), meaning that the entirety of Chapter 2 is now redundant, and what I say in that chapter about giraffe neck anatomy and evolution has been challenged by research published in 2007 and 2009 (Cameron & du Toit 2007, Badlangana *et al.* 2009, Mitchell *et al.* 2009). The three living manatee species are discussed in Chapter 3. Since that article was written, a fourth living manatee species has been proposed: the alleged dwarf, Amazonian species *Trichechus pygmaeus*. The article naming this species does not yet exist in print (it was published online by primatologist Marc van Roosmalen in 2008), and at the moment it does not appear likely that *T. pygmaeus* will be accepted as valid by other mammalogists (sirenian expert Daryl Domning is on record as stating that it is based on juvenile specimens of *T. inunguis*).

Reticulated pythons are discussed in Chapter 5. This spectacular snake is most usually known as *Python reticulatus* but a new generic name, *Broghammerus*, was proposed for the species in 2004 (Hoser 2004). While this proposal has not been widely accepted (Raymond Hoser is, to put it mildly, a controversial figure in herpetological systematics), the distinction of *Broghammerus* relative to *Python* and other python taxa has been supported in some other studies. The

2008 publication of *Biology of the Snapping Turtle* (Steyermark *et al.* 2008) – a massive, multi-authored volume containing stacks of information on snapping turtles – means that Chapter 6 would be very different if written today. Chapter 9 – focusing on early cetaceans – might also be somewhat different if written today, given that so much has happened in the field of Eocene cetaceans and cetacean origins since 2006. How whales are related to other hoofed mammals has now become one of the biggest questions in palaeomammalogy. Several major studies have appeared where Eocene whales and a large number of archaic, fossil hoofed mammals have appeared in the same analysis. The closest *living* relatives of whales might be hippos, but several or many fossil groups may have been even closer.

Chapter 10 discusses some of the evidence that might be used to support the existence of British big cats, and leans heavily on a Roe deer carcass discovered at Cupar in Scotland. Since I wrote the original Chapter 10 text (in August 2006) I've become acquainted with the large amount of field data collected by Jonathan McGowan of the Bournemouth Natural Science Society. Jon's data (which includes photographic documentation as well as the remains of carcasses) means that the Cupar carcass is far from unique, though the point remains that it's one of the most convincing and best documented. Compelling field evidence for the reality of British big cats has also since been published by Coard (2007). Moving to something entirely different, Chapter 11 refers to on-going work on British sauropod dinosaurs, and mentions what was (at the time) a newly recognised, unnamed taxon. That taxon is *Xenoposeidon proneneukos*, and it was published by Taylor & Naish (2007). Also relevant is that Naish & Martill (2007) have since published a review of British sauropods. Also on sauropods, Chapter 12 refers to the discovery of rebbachisaurid sauropods in the Lower Cretaceous of England. Additional rebbachisaurid elements from England have since been published by Mannion (2008), and it should also be noted that I'd overlooked the somewhat older European rebbachisaurid remains from Croatia (Dalla Vecchia 1998).

Chapter 22, a discussion of azhdarchid pterosaur lifestyle and palaeobiology, finishes by noting that Mark Witton and me will one day produce a paper on this very subject. Some time after the Tet Zoo article was published, Mark and I got to work on this project, and our technical, peer-reviewed paper on the subject appeared in 2008 (Witton & Naish 2008). It received a huge amount of interest and is one of the most discussed, most popular academic projects I've been involved in. This paper, and a 2008 Tet Zoo article I wrote to accompany it, substantially updates the 2006 article included in this book.

Several new publications have appeared since 2006 on the subject of bird predation in Greater noctules (a subject covered in Chapter 38). The Isturitz statuette – stated in Chapter 17 to represent the sabretoothed cat *Homotherium latidens* – has now been argued to be a representation of a Cave lion *Panthera leo spelaea* (Antón et al. 2009).

Chapter 29 discusses the taxonomy of the legless anguid lizards included within the genus *Anguis* – these lizards are popularly known as slow-worms – and says that there are two recognised species: *A. fragilis* and *A. cephallonica.* Gvoždík *et al.* (2010) have since presented genetic data which indicates that *A. fragilis* of tradition should be split up into three species: in addition to *A. fragilis* (Common European slow-worm) of northern Europe, they recognise *A.*

colchica (Eastern slow-worm) of southern Russia and western Georgia, and *A. graeca* (Greek slow-worm) of Greece. If their taxonomic proposals are eventually accepted, there are now four slow-worm species.

As the palaeanodonts, pangolins, fairy armadillos and marsupial moles looked on, all Johnny the amphisbaenian could say was....

Merry Christmas!

Best fossorial wishes for Christmas and New Year 2005/6, from Darren Naish

Fig. 2. The Christmas card that started it all, sort of. Four years later and I still haven't written about palaeanodonts, pangolins or amphisbaenians.

Refs - -

- Antón, M., Salesa, M. J., Turner, A., Galobart, Á. & Pastor, J. F. 2009. Soft tissue reconstruction of *Homotherium latidens* (Mammalia, Carnivora, Felidae). Implications for the possibility of representations in Palaeolithic art. *Geobios* 42, 541-551.
- Badlangana, N. L., Adams, J. W. & Manger, P. R. 2009. The giraffe (*Giraffa camelopardalis*) cervical vertebral column: a heuristic example in understanding evolutionary processes? *Zoological Journal of the Linnean Society* 155, 736-757.
- Berger, L. R. 2006. Predatory bird damage to the Taung type-skull of *Australopithecus africanus* Dart 1925. *American Journal of Physical Anthropology* 131, 166-168.
- Berger, L. R. & McGraw, W. S. 2003. Further evidence for eagle predation of, and feeding damage on, the Taung child. *South African Journal of Science* 103, 496-498.

- Brown, D. M., Brenneman, R. A., Koepfli, K.-P., Pollinger, J. P., Milá, B., Georgiadis, N. J., Louis, E. E., Grether, G. F., Jacobs, D. K. & Wayne, R. K. 2007. Extensive population genetic structure in the giraffe. *BMC Biology* 2007, 5: 57 doi:10.1186/1741-7007-5-57
- Cameron, E. Z. & du Toit, J. T. 2007. Winning by a neck: tall giraffes avoid competing with shorter browsers. *The American Naturalist* 169, 130-135.
- Coard, R. 2007. Ascertaining an agent: using tooth pit data to determine the carnivores responsible for predation in cases of suspected big cat kills. *Journal of Archaeological Science* 34, 1677-1684.
- Dalla Vecchia, F. M. 1998. Remains of Sauropoda (Reptilia, Saurischia) in the Lower Cretaceous (upper Hauterivian/lower Barremian) limestones of SW Istria (Croatia). *Geologia Croatica* 51, 105-134.
- Gvoždík, V., Jandzik, D., Lymberakis, P., Jablonski, D. & Moravec, J. 2010. Slow worm, *Anguis fragilis* (Reptilia: Anguidae) as a species complex: genetic structure reveals deep divergences. *Molecular Phylogenetics and Evolution* 55, 460-472.
- Hoser, R. 2004. A reclassification of the Pythoninae including the descriptions of two new genera, two new species, and nine new subspecies. Part II. *Crocodilian* 4, 21-40.
- Mannion, P. 2008. A rebbachisaurid sauropod from the Lower Cretaceous of the Isle of Wight, England. *Cretaceous Research* 30, 521-526.
- Mitchell, G., van Sittert, S. J. & Skinner, J. D. 2009. Sexual selection is not the origin of long necks in giraffes. *Journal of Zoology* 278, 281-286.
- Naish, D. & Martill, D. M. 2007. Dinosaurs of Great Britain and the role of the Geological Society of London in their discovery: basal Dinosauria and Saurischia. *Journal of the Geological Society, London* 164, 493-510.
- Steyermark, A. C., Finkler, M. S. & Brooks, R. J. 2008. *Biology of the Snapping Turtle* (Chelydra serpentina). The John Hopkins University Press (Baltimore).
- Taylor, M. P. & Naish, D. 2007. An unusual new neosauropod dinosaur from the Lower Cretaceous Hastings Beds Group of East Sussex, England. *Palaeontology* 50, 1547-1564.
- Witton, M. P. & Naish, D. 2008. A reappraisal of azhdarchid pterosaur functional morphology and paleoecology. *PLoS ONE* 3 (5): e2271. doi:10.1371/journal.pone.0002271

Acknowledgements

The acknowledgements section of a book is, ordinarily, a list of thanks offered to those people who assisted the author during the production of the book. Writing the acknowledgements for a book based on a blog is different though, because it only seems right to thank the people who helped with the blog. That list is pretty long, but… what the hell.

First of all, and above all others, I must thank Toni for her unbelievable tolerance and support. My writing and research does infringe substantially into my social and family life, and I'm incredibly lucky to have such an understanding wife. Moving on, thanks to Jon Downes and everyone else at the CFZ for seeing this book through to production, and thanks to Jon for suggesting I put the book together in the first place. Thanks also to Karl Shuker for his advice and encouragement on the project. I also want to thank up-front those who have been real Tet Zoo champions since the early days, and who provided ego-boosting praise, moral support, and invaluable assistance in obtaining literature.

I thank Mathew Wedel and Mike P. Taylor, Steve Bodio and friends at Querencia, the omniscient Randy Irmis, and Dave Hone, who I think I only got to know *because of* Tet Zoo. Mark Witton, Richard Hing and Graeme Elliott of the Portsmouth Palaeobiology Research Group provided endless Three-Stooges-style hilarity, but also assisted in trips to the zoo and in insightful critique. Mark is also warmly thanked for allowing the use of several photos and pieces of art, one of which (Fig. 64) appears here for the first time ever. Dave Martill had a vested interest in me *not* writing Tet Zoo, but came round eventually and might even have visited the blog once or twice. Yet another Dave, Herr Unwin, deserves thanks for his support. Stig Walsh provided helpful information on giraffes, seals and other subjects, and I used to like telling people that he was the Paul Burrell to my Princess Diana (though this unusual choice of description is looking increasingly problematical and I should stop using it). Phil Budd and Bernard Dempsey assisted me on various adventures involving British wildlife, and Sarah Fielding provided the inspiration for various turtle-based side-projects that evolved into blog posts. Tommy Tyrberg provided helpful data on birds, and useful opinions on island-dwelling tortoises and crocodilians were provided by Julian Hume.

Marc van Roosmalen, Carl Buell, Luis Rey, Ben Speers-Roesch, C. M. Koseman and John Conway are thanked for discussion, opinion and co-operation. Neil Phillips kindly allowed use of several of his photographs. I also thank the several individuals whose photos or artwork appear here courtesy of wikipedia and other online sources, including Franco Andreone, Dmitry Bogdanov, Thierry Caro, John Hill, Arne Hodalic, B. S. Thurner Hof, Evgenia Kononova, Adam Kumiszcza, Matt Martyniuk, Dave Pape, Burkhard Plache, Ryan E. Poplin, Alastair Rae, Portia Sloan, Ryan Somma, Josiah H. Townsend, David Vieites and Arthur Weasley.

The photos of the Cupar roe deer carcass appear courtesy of Ralph Barnett and Big Cats in Britain; many thanks to Mark Fraser for procuring permission. C. M. Koseman is sincerely thanked for producing the excellent cover art at very short notice and deserves extraordinary praise.

Many other people who visit or read Tet Zoo were kind enough to allow me to use photographs or illustrations: sincere thanks to them all. If there's anyone I've neglected to thank or mention please accept my apologies! Lastly, I thank the many people who read Tet Zoo and have provided helpful comments, feedback and support. There is still a heck of a lot of Tet Zoo to get through before my work is done...

Chapter 1:
When eagles go bad

One of my favourite 'fringe' subjects within tetrapod zoology concerns the alleged ability of eagles to attack and kill unusually large mammals, including people. Most people, and indeed most ornithologists and other zoologists, don't take this notion too seriously, or indeed dismiss it immediately as utter nonsense. Fortean literature is replete with stories of eagles (usually European Golden eagles *Aquila chrysaetos*) attacking, killing and/or carrying off people, typically (but not always) young children, with the best known story of this sort being that of 5-yr-old Marie Delex from the French Alps. It's pretty inconceivable that even the biggest eagle could carry off a human child, but as for whether they could kill one, well, read on.

Two recent items in the media have concerned this issue, and both caught my attention.

Fig. 3. Giant eagles such as the hypothetical species shown here once posed a very real danger to our ancestors, the australopithecines.

Firstly, we have news from researchers at Ohio State University on the death of the Taung child (or Taung baby, depending on your preference). This is the famous juvenile australopithecine specimen described by Dart in 1925, and which was thought to have been a 3 or 4 year old. Following up on C. K. Brain's observations of 1981 that the Taung assemblage represented an accumulation produced by a large carnivore, probably a leopard, Lee R. Berger and Ronald J. Clarke (1995) showed in the *Journal of Human Evolution* that a large eagle was the most likely killer of the juvenile. The case was good: the assemblage consists of smallish mammals (like mole rats, spring hares and small antelopes), evidence for carnivorous mammals is absent, nick marks corresponding to those produced by eagle beaks and talons are present on some of the bones, and eggshell was also discovered at the site. The new discovery is that nick marks around the orbital margins of the Taung child demonstrate once and for all that an eagle really was the killer. Great stuff – I look forward to the paper.

What's a little odd is that several people have started asking questions about the lifting abilities of the Taung eagle relative to the weight of the juvenile australopithecine (oh, and by the way, the eagle is [I understand] presently a theoretical one (Fig. 3) – the evidence for its presence is there, but the eagle itself has yet to be found. It has been considered by some workers that the African crowned eagle *Stephanoaetus coronatus* was the most likely culprit – more on that at another time). This is odd because it was discussed to death the last time there was a flurry of interest in the 'eagle as killer' theory. Anders Hedenström (1995) showed in *Nature* that, given that *Stephanoaetus* can carry animals weighing just over 6 kg (wow!), then it would have to have dismembered the australopithecine. Which is fine, given that the specimen is only known from its skull. However, Hedenström argued this based on an assumed mass for the Taung child of about 10 kg, and that might have been too high.

Berger & Clarke (1996) countered that evisceration of the juvenile prior to its carrying was likely, given that large extant raptors commonly do this. This could mean a 30% loss in body weight of the juvenile, thus bringing it close to or within the short-range carrying abilities of *Stephanoaetus*. It's also worthy of note that Berger and Clarke brought attention to cases where big living African eagles simply must have lifted animals that weighed more than 6 kg, so this perhaps wasn't the 'upper' lifting limit that Hedenström thought it was.

So, to those people who have suggested that the Taung eagle killed the australopithecine there on the spot (oh, I see, coincidentally in the middle of a veritable midden pile of eagle-killed small mammals), I say check out the literature on eagle lifting abilities.

Speaking of killing the australopithecine on the spot, this introduces the next thing I wanted to talk about... can a big eagle kill a mammal that's too big for the eagle to carry? Like it or not – and I've received some vitriolic emails for promoting this idea (Naish 1998, 1999) – the answer is a definite yes. In fact the ability of large eagles, specifically Golden eagles, to kill relatively enormous prey is not doubted and well established. Between 1987 and 1989 representatives of Animal Damage Control (ADC) were called in to check out mysterious domestic cattle deaths that were occurring in Socorro County, New Mexico. 6 calves were killed and 13 injured, with the biggest calf attacked weighing 115 kg (Fig. 4). The attacks were caused, unquestionably, by a pair of local Golden eagles, as verified by observed attacks and by the talon marks on calf skulls. The problem stopped when the eagles were removed. It seems the eagles killed by puncturing the skull base. I know this all sounds pretty incredible, ridiculous even: if you don't believe me see Robert Phillips *et al.*'s 1996 paper in *Wildlife Society Bulletin*. You might be as surprised as I was to learn that Golden eagles are also documented as killers of Mule deer, Pronghorn and semi-domestic reindeer, and again this is all documented in the technical literature and I'm not making it up. For killing of juvenile deer, see Cooper (1969) for Red deer and Ratcliffe & Rowe (1979) for Roe deer.

Fig. 4. The Golden eagle can be a macropredator, well able to kill giant mammalian prey.

What was the second thing in the media that I wanted to mention? On Jan' 17th 2006, the BBC broadcast 'Bill Oddie's How to Watch Wild-life' (or whatever), and it included something I've never seen before on film. Oddie and a col-league were observing a wild Golden eagle in the Cairngorms, and the camera was on the bird too. Suddendly, it swooped down low on an adult Red deer.

The deer ran down the hillside, pursued closely by the eagle, which swerved and jinked to follow and harass it. The eagle wasn't about to dig its talons into the deer; rather, it looked as if it was trying to see if it could get the deer to take a nasty fall down the hill. This behaviour has been mentioned for some other raptors – Lammergeiers *Gypaetus barbatus* are said to do it to ibex and chamois – but I didn't know Golden eagles did it.

On Lammergeiers, Berridge (1934) wrote "A favourite method of dealing with [ibex and chamois] is to swoop down suddenly upon a prospective victim that may be poised somewhat insecurely upon the steep hillside, so that the startled beast loses its foot-hold, and goes tum-bling to death in the ravine below" (p. 219). The scoop is that Lammergeiers don't just try this out on ibex and chamois, they will also try this on humans too, and I know this because it has been reliably reported by a professional biologist (I don't have his permission to cite it as a pers. comm., but will try and get this on record).

One final thought on this subject, though to be honest it's not that original and many people end their articles on nasty eagles with the same thought. The raptors I've been discussing here aren't particularly big: Golden eagles max out at 6.6 kg, and *Stephanoaetus* is less than that. Given that there were, recently, extinct eagles that were substantially bigger than this – Haast's eagle from New Zealand weighed about 13 kg (in females) for example – then eagles as a group simply must have been capable of even more amazing feats of predation.

Haast's eagle

Having mentioned Haast's eagle, it's worth looking at this species in more detail. Until re-cently New Zealand truly was a land of birds. Inhabited by the 11 (or so) species of moa, the long-beaked kiwis, snipes and snipe-rails, the bizarre adzebills, a variety of rails and coots, flightless ducks and geese, giant terrestrial owlet-nightjars, diminutive New Zealand wrens, and a motley assortment of crows, quails, mergansers, parrots, wattlebirds, thrush-like passer-ines, owls, honeyeaters and herons, it would appear to be the ideal place for birds of prey to evolve, and to take predatory advantage of this diverse avifauna. We now know that New Zea-land was inhabited until very recent times by at least two such birds of prey, and both/all were specialised bird predators [an alleged third species, a sea eagle described from Chatham Island in 1973 and dubbed *Haliaeetus australis*, has proved to be a mis-labelled North American Bald eagle *H. leucocephalus*. A fourth species, the New Zealand falcon *Falco novaeseelandiae*, has survived to the present].

The first of the two bird predators was originally described as a very large harrier, and called *Circus eylesi* by Ron Scarlett in 1953. With large females perhaps weighing as much as 3 kg, this was a giant if it were a harrier: living species rarely weigh more than 700 g (Clarke 1990). However, Scarlett later regarded the bird as a giant goshawk (a member of the genus *Accipiter*): a specialist bird-catcher adept at flying through tangled woodland habitats. Though the goshawk reidentification has become quite well known, new examination has demonstrated that the attribution of the species to *Circus* was correct. It really was a gigantic, bird-killing member of the group (and whether it is one or two species still remains contested).

The second New Zealand bird of prey was the biggest eagle EVER: the enormous Haast's eagle *Harpagornis moorei* Haast 1872, a powerful forest giant that some experts have imagined as being something like the modern Harpy eagle *Harpia harpyja*. As is discussed below, new data indicates that the generic name *Harpagornis* is invalid, but we won't worry about that now.

Introducing Haast's eagle

Haast's eagle has been known to science since 1871, but until recently virtually nothing was known of how it may have lived, or how it was related to other kinds of eagles. Several publications on Haast's eagle addressing these problems were produced by Richard Holdaway in the late 1980s and 1990s following his 3-volume doctoral dissertation, and a huge amount of information on the bird was recently compiled by Trevor Worthy and Holdaway for their superb book *The Lost World of the Moa*.

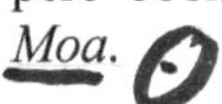

The first Haast's eagle material to be discovered by Europeans was found by Frederick Fuller at the moa excavation site at Glenmark

Fig. 5. In this reconstruction, produced by John Megahan and originally appearing in Bunce *et al.* (2005) in *PLoS Biology*, Haast's eagle attacks a moa. Note that – in contrast to some older reconstructions – Megahan's version of Haast's eagle much resembles the living eagles of the genera *Aquila* and *Hieraaetus*.

swamp, Canterbury, in 1871. Johann Franz Julius Haast (1822-1887), director of the Canterbury Museum and noted expert on moa and other New Zealand birds, read a description of the species to the Philosophical Society of Canterbury in 1871 and published his description the following year (Worthy & Holdaway 2002). He named the bird after George Moore, the owner of Glenmark Station. This association of Haast's eagle with moa immediately led to suggestions that it was a moa-hunter, and had perhaps come to feed on moa that had become trapped in the mud of the swamps.

Subsequent finds showed the eagle to be widely distributed on South Island and the southern half of North Island. However, even before its official extinction date, it seems to have been rare or absent from the eastern coast of central North Island (Horn 1983). Quite why it was never widely distributed in the northern half of North Island, which at this time was heavily forested and apparently ideal for the bird, remains a mystery.

Like other giant birds of prey around the world, Haast's eagle does not appear to have been common at any locality and its remains are not particularly abundant: only three complete skeletons are known, the most recently discovered being reported in 1990. This specimen was found at the bottom of a narrow vertical sinkhole and seems to have fallen accidentally to its death (Worthy & Holdaway 2002).

As in most birds of prey, female Haast's eagles were larger than males. In estimated weight they were 10-13 kg and, though their wings were proportionally short (in keeping with its probable forest-dwelling lifestyle), they still had a wingspan of around 2.6 m.

The standing height of a female Haast's eagle has been estimated as 1.1 m. Male Haast's eagles, which were initially regarded as a separate species (*H. assimilis* Haast, 1874), probably weighed 9-10 kg. For comparison, the largest living eagles, the Harpy *Harpia harpyja*, Philippine eagle *Pithecophaga jefferyi* and European *Haliaeetus albicilla* and Steller's sea eagles *H. pelagicus*, rarely exceed 2.4 m in wingspan. Some female *H. albicilla* have been recorded with wingspans of 2.65 m and a large female *H. pelagicus* may reach 9 kg (Brown 1976, Burton 1989). Amongst living birds of prey, only condors exceed these measurements: the Andean condor *Vultur gryphus* possibly exceeding 3 m in wingspan and reaching 12 kg.

The skull of Haast's eagle measures 15 cm in total length and is elongate, looking something like a stretched version of an *Aquila* skull, and without the tremendously deep beak seen in some forest eagles like *Pithecophaga* and *Harpia*. The skull is actually superficially rather like that of an Old World vulture and, unlike other aquiline eagles, the nostril in Haast's eagle is partially closed-off by ossification around its rostral, dorsal and ventral margins. This recalls the presence of an accessory bony plate that covers part of the nostril opening in some of the larger Old World vultures.

The legs and feet of Haast's eagle are tremendously robust and powerful and appear well capable of dispatching very large prey. Its claws are massive, strongly curved and, with their external keratin sheath, would have reached 75 mm in length. A complete, thorough description of the osteology of Haast's eagle was provided by Worthy & Holdaway (2002).

Fig. 6. Another (somewhat older) reconstruction of Haast's eagle, again shown attacking an unfortunate moa. As is typical of older reconstructions, this one makes the eagle look like a Philippine or Harpy eagle.

While some material dates Haast's eagle to around 30,000 years before present, its youngest remains show that it was still around about 500 years ago, and it therefore most probably became extinct at around the same time as (or slightly before) the last moa. Presumably, as moa hunting became more intense and moa became rarer and rarer, and as habitat degradation on New Zealand increased, Haast's eagle became increasingly pressurised and eventually unable to sustain a population. A c. 300 year overlap of Haast's eagle and humans therefore seems to have occurred (the Maori colonised New Zealand at around 850 years before present).

Seeing as the Maori have legends of giant predatory birds – including of the sky-dwelling hokioi or hakuwai and the man-eating pou-kai – it has been tempting for writers to speculate that these legends do indeed refer to Haast's eagle (Reed 1963, Hall 1994). Unfortunately the legends are far too vague for this to be confirmed, but it has led to some very interesting speculations.

Surprisingly, much as some people have continued to entertain notions of moa survival to the present day, there have been a number of suggestions that Haast's eagle did not die out 500 years ago, but survived to within living memory. This idea is based on two lines of evidence: eyewitness accounts of large birds of prey, and reports of unusual unidentified bird calls of a particularly loud and startling nature. The calls were reported from Stewart Island as recently as 1961 and, because the animal making these nocturnal sounds was never seen, it was suggested by some that Haast's eagle might have been the vocaliser. Needless to say, more likely candidates are on offer and Miskelly (1987) suggested that New Zealand snipe (*Coenocorypha*) might actually be the culprit.

Alleged eyewitness accounts from recent times are few, and restricted to the 1860s. Haast actually saw what he thought was a giant eagle in the Canterbury mountains, and a large bird that walked into his tent one night was also suggested to be a giant eagle. Even more incredible is Charlie Douglas's reported shooting of two giant eagles in the Landsborough Valley, both of which, reportedly, had wingspans of 3 m or so. Worthy & Holdaway (2002, p. 335) noted that Douglas was "a meticulous observer, he did not seek publicity, and he certainly never claimed that he had seen the "extinct" eagle". It's an incredible report, and one that can't be verified.

The origin and evolution of Haast's eagle

Back in 2005 the big deal was that Haast's eagle – traditionally given its own genus (*Harpagornis*) and thought to be close to or part of the big-bodied *Aquila* eagle clade (Holdaway 1994) – turned out to be part of a clade of small-bodied eagles. This discovery means that we have to revise our ideas on the affinities of the giant eagle, but it also shows that a totally unexpected and incredible thing happened during the evolution of this bird's lineage.

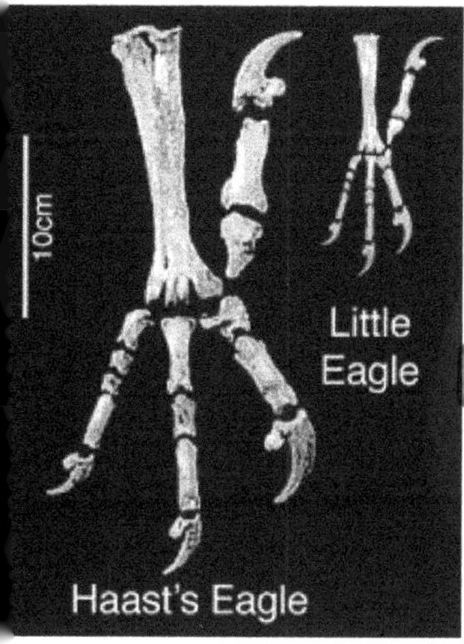

Fig. 7. The enormous foot of a Haast's eagle compared with that of one of its closest relatives, the much smaller Australian little eagle *Hieraaetus morphnoides* (it is now argued by some that the *Hieraaetus* eagles should be included within the genus *Aquila,* in which case the Australian little eagle is *Aquila morphnoides*). This image originally appeared in Bunce *et al.* (2005) in *PLoS Biology*

As demonstrated by Bunce *et al.* (2005), DNA from Haast's eagle indicates that the species is actually nested within *Hieraaetus*, the hawk-eagles, and hence should be renamed *Hieraaetus moorei*. All other *Hieraaetus* species are quite small (they weigh about 1 kg and have wingspans of c. 1.2 m), so the ancestor of Haast's eagle must have been small too. It is inferred that, on arriving on a landmass lacking competitors and with an abundance of large, terrestrial prey, a small *Hieraaetus* species rapidly evolved into a powerful giant. We know from well documented taphonomic evidence that Haast's eagle could and did kill even the very biggest of the contemporary moa, and – unlike big eagles in continental environments – it could probably get away with gorging itself at a carcass without having to worry about flying off in a hurry.

Refs - -

- Berger, L. R. & Clarke, R. J. 1995. Eagle involvement in accumulation of the Taung child fauna. *Journal of Human Evolution* 29, 275-299.
- Berger, L. R. & Clarke, R. J. 1996. The load of the Taung child. *Nature* 379, 778-779
- Berridge, W. S. 1934. *All About Birds*. George G. Harrap, London.
- Brown, L. 1976. *Eagles of the World.* David & Charles, Newton Abbot/London/ Vancouver.
- Bunce, M., Szulkin, M., Lerner, H. R. L., Barnes, I., Shapiro, B., Cooper, A. & Holdaway, R. N. 2005. Ancient DNA provides new insights into the evolutionary history of New Zealand's extinct giant eagle. *PLoS Biology* 3 (1), e9 DOI: 10.1371/ journal.pbio.0030009
- Burton, P. 1989. *Birds of Prey*. Gallery Books, London.
- Clarke, R. 1990. *Harriers of the British Isles.* Shire Natural History, Princes Risbor-

ough, UK.

- Cooper, A. B. 1969. Golden eagle killed Red deer calf. *Journal of Zoology* 158, 215-216.
- Hall, M. A. 1994. *Thunderbirds - the living legend!* (2nd edition). Privately published, Minneapolis.
- Hedenström, A. 1995. Lifting the Taung child. *Nature* 378, 670.
- Holdaway, R. N. 1994. An exploratory phylogenetic analysis of the genera of the Accipitridae, with notes on the biogeography of the family. In Meyburg, B.-U. & Chancellor, R. D. (eds) *Raptor Conservation Today*. WWGBP/The Pica Press, pp. 601-649.
- Horn, P. L. 1983. Subfossil avian deposits from Poukawa, Hawkes Bay, and the first record of *Oxyura australis* (Blue-billed duck) from New Zealand. *Journal of the Royal Society of New Zealand* 13, 67-78.
- Miskelly, C. M. 1987. The identity of the hakawai. *Notornis* 34, 95-116.
- Naish, D. 1998. Big bad eagles 2: sheep and cow on the Golden eagle menu. *Mainly About Animals* 36, 10-12.
- Naish, D. 1999. Big bad killer eagles. *Fortean Times* 122, 48.
- Phillips, R. L., Cummings, J. L., Notah, G. & Mullis, C. 1996. Golden eagle predation on domestic calves. *Wildlife Society Bulletin* 24, 468-470.
- Ratcliffe, P. R. & Rowe, J. J. 1979. A Golden eagle (*Aquila chrysaetos*) kills an infant Roe deer (*Capreolus capreolus*). *Journal of Zoology* 189, 532-535.
- Reed, A. W. 1963. *Treasury of Maori Folklore*. A. H. & A. W. Reed, Wellington, New Zealand.
- Worthy, T. H. & Holdaway, T. H. 2002. *The Lost World of the Moa*. Indiana University Press, Bloomington, Indiana.

Chapter 2:
Giraffes: set for change

Giraffes, *Giraffa camelopardalis*. Everybody loves giraffes – how could you not. Giraffes are such fantastic and unusual mammals that all manner of questions are still being asked about their anatomy, physiology and behaviour. A mild amount of controversy exists over how many neck vertebrae they have: the old chestnut about them having the same number of cervical vertebrae as most other mammals (seven) has been challenged by Solounias (1999). Why they have long necks remains contentious – personally I've become a fan of the sexual selection hypothesis championed by Simmons & Scheepers (1996) – and basic questions of anatomy, such as how their circulation works (Pedley *et al.* 1996) and why their bones differ in density from those of other artiodactyls (van Schalkwyk *et al.* 2004), remain interesting areas of research.

As the only extant long-necked terrestrial tetrapods, giraffes are much employed in debates over the neck posture of sauropod dinosaurs, and recent articles by John Martin and colleagues (Martin *et al.* 1998), and Kent Stevens and J. M. Parrish (2005), discuss giraffes and what they might tell us about sauropod necks. Greg Paul has recently been saying a lot about giraffe necks on the dinosaur mailing list, but I haven't yet read his comments, and I know at least two sauropod workers who have something in the pipeline on this subject. Something is happening in the world of giraffe research that will probably be unknown to dinosaur workers, and will affect the way in which giraffes are named in the literature. It's no big deal, but it's worth bringing to attention: it's the species level taxonomy. Pretty soon, it's no longer going to be ok to talk of your giraffe as being (necessarily) a *Giraffa camelopardalis*. Why? Because there are studies underway which indicate that more than one extant species is present.

Nine *Giraffa camelopardalis* subspecies are currently recognized, and while some authors have been happy to accept all of them as valid, others have wondered whether at least some of the variation is just individual. Indeed there are photos showing individuals of different supposed 'subspecies' standing next to each other (Dagg 1962).

Fig. 8. A selection of giraffes photographed at a zoo: these are Reticulated giraffes. Theropod expert Nick Longrich stands in the foreground.

For the record, the subspecies are *G. c. angolensis* (Angolan giraffe), *G. c. antiquorum* (Kordofan giraffe), *G. c. camelopardalis* (Nubian giraffe), *G. c. giraffa* (Southern/South African giraffe), *G. c. peralta* (Nigerian/West African giraffe), *G. c. reticulata* (Reticulated giraffe), *G. c. rothschildi* (Baringo/Rothschild's/Uganda giraffe), *G. c. thornicrofti* (Thornicroft's giraffe) and *G. c. tippelskirchi* (Masai giraffe)

At least some of the observed variation is real (and not just individual variation, or clinal) – but how much of it, and what does it mean for taxonomy and phylogeny? Russell Seymour has recently been giving talks with titles like 'How many giraffes? Temporo-spatial evolution in a "well-known" species and its implications for conservation of biodiversity', so, reading between the lines, there's the implication here that *Giraffa camelopardalis* is actually more than one species. Maybe one or more of the supposed subspecies is going to be elevated to species rank – or in fact *restored* to species rank, because at least some of the taxa were initially named as species, but later demoted when it became fashionable to lump big mammals in this way (and there's an essay in that subject, by the way). A team involving Seymour, Rick Brenneman, David Brown, Thomas deMaar and Julian Fennessy are studying giraffe phylogeography, and their results should have bearing on this matter. Seymour has also been listed alongside Norman MacLeod and M. Bruford as looking at variation within *Giraffa camelopardalis*. I've seen Norman MacLeod listed as a coauthor on presentations regarding DAISY, an image-recognition software programme being developed at the Natural History Museum in London. Basically, they've developed software that can 'look at' and identify objects (including microfossils and bird bones) and, reading between the lines again, I wonder if it's been applied to the dis-

Fig. 9. Another captive Reticulated giraffe. Photo by Neil Phillips.

tinctively patterned coats of giraffe subspecies? We'll find out in due course. 

So far as I know nothing has yet been published on this – please let me know if you know otherwise. But, to those who are planning to mention giraffes in any of their writings, do keep this in mind: the giraffe you are thinking of may end up being something other than *Giraffa camelopardalis*.

Refs - -

- Dagg, A. I. 1962. The subspeciation of the giraffe. *Journal of Mammalogy* 43, 550-552.
- Martin, J., Martin-Rolland, V. & Frey, E. 1998. Not cranes or masts, but beams: the biomechanics of sauropod necks. *Oryctos* 1, 113-120.
- Pedley, T. J., Brook, B. S. & Seymour, R. S. 1996. Blood pressure and flow rate in the giraffe jugular vein. *Philosophical Transactions of the Royal Society of* London *B* 351, 855-866.
- Simmons, R. E. & Scheepers, L. 1996. Winning by a neck: sexual selection in the evolution of giraffe. *The American Naturalist* 148, 771-786.
- Solounias, N. 1999. The remarkable anatomy of the giraffe's neck. *Journal of Zoology* 247, 257-268.
- Stevens, K. A. & Parrish, J. M. 2005. Neck posture, dentition and feeding strategies in Jurassic sauropod dinosaurs. In Tidwell, V. & Carpenter, K. (eds) *Thunder-Lizards: The Sauropodomorph Dinosaurs*. Indiana University Press (Bloomington & Indianapolis), pp. 212-232.
- van Schalkwyk, O. L., Skinner, J. D. & Mitchell, G. 2004. A comparison of the bone density and morphology of giraffe (*Giraffa camelopardalis*) and buffalo (*Syncerus caffer*) skeletons. *Journal of Zoology* 264, 307-315.

Chapter 3:
Those transatlantic manatees

There are three extant manatee species: *Trichechus inunguis* of the Amazon Basin, *T. manatus* of the Gulf of Mexico, Caribbean and US Atlantic coast as far north as Virginia, and *T. senegalensis* of western Africa (Fig. 11). So, how it is that they occur on opposite sides of the Atlantic? A very old idea explains it thus: as the Americas and the Old World rifted apart in distant geological times, the ancestral manatee species got separated and, presto, a vicariance event resulted in speciation. However, I'm sure I don't need to tell you how absurd it is, today, to suggest this. The North Atlantic opened something like 100 million years ago, yet manatees (well, those of the extant genus *Trichechus* anyway) are probably less than 10 my old (there being questionable *Trichechus* fossils from the Pliocene). Granted, there are sirenian workers who have, indeed, suggested that Atlantic rifting might explain manatee distribution… but, those workers were publishing their papers in the early years of the 20th century (Arldt 1907). Dispersal is clearly the only option: that is, yes, manatees simply must have crossed the Atlantic at some stage, and a quick check of the literature on manatee evolution reveals many references to this hypothesis.

Based on a spurious idea about North Atlantic currents, Simpson (1932) thought that manatees migrated from east to west. However, the evidence clearly shows west to east to be more likely; the fossil trichechine phylogenetically closest to *Trichechus* (Mio-Pliocene *Ribodon*) is South American, and in fact all fossil trichechines are American; *T. manatus* and *T. senegalensis* are more like each other than either is to *T. inunguis*; the nematode parasites of *T. senegalensis* seem to be more specialised than the nematodes of *T. manatus*, and so on.

So, given that manatees simply must have crossed the Atlantic to get to Africa, how did they do it? Well, they swam of course, and the really cool thing is that there are also reasons for thinking that this isn't such a big deal: it's plausible, and in fact it's supported by strange things that manatees have done in historical times (and I'll return to that in a minute). Daryl Domning, world expert on sirenian evolution and history, published a paper on manatee evolu-

Fig. 10. West Indian manatee *Trichechus manatus*. This one was photographed in Florida, not in the middle of the Atlantic.

tion in *Journal of Vertebrate Paleontology* last year. In explaining the successful invasion of the African coast by American manatees, he brought in all of the arguments given above, but tied it together with data on Amazonian and Atlantic palaeocurrents. During the Pleistocene, the subtropical North Atlantic gyre was compressed, and a cold current ran along the Eurafrican coast to as far south as Gambia. Boekschoten & Best (1988) explained how this appears to have allowed Caribbean corals and certain molluscs to colonise the eastern Atlantic, and they also speculated that manatees also used this route. Furthermore, it turns out that an 'appreciable fraction' of Amazon River water gets right across the Atlantic as far as Africa, so "manatees taking this route might even have access to relatively fresh water for a good part of the journey, if they rode in a large lens of Amazon water" (Domning 2005, p. 699).

And the trump card? Domning (2005) suggested that purported manatee strandings made on the coasts of the North Atlantic in historical times may really have been genuine. Animals alleged to have been manatees have been reported from the shores of Greenland (1780), Scotland (1801 and 1837) and France (1782).Whilst we should remain sceptical about these accounts, it's not implausible that they were genuine. Domning cites a radio-tracked Florida manatee that, in 1995, got as far as Rhode Island.

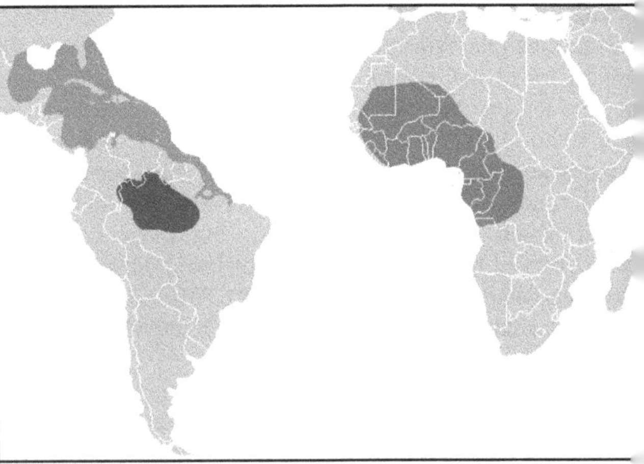

Fig. 11. The distribution of the three living manatee species. In terms of distribution, the African species is the weird one.

Here's another spin on this. A long-standing mystery in the cryptozoological literature has been the purported presence of manatees on St. Helena in the South Atlantic (not too far south: St. Helena is at the same latitude as Bolivia, Angola, or northern Madagascar). Nobody's ever really known what these animals were – were they really manatees, or were they actually seals of some kind? The several descriptions provided sound to me like those of pinnipeds (mostly sea lions), and indeed most authors have concluded that this is what the animals were. After reviewing the mystery, Shuker (1995) left the case open, however. While, previously, there were good reasons for doubting the idea that manatees might ever have gotten to St. Helena, our new understanding of manatee dispersal at least renders this idea a remote possibility. In other words, it probably is just about conceivable that manatees could have gotten to St. Helena after all.

Refs - -

- Arldt, T. 1907. Zur Atlantisfrage. *Naturwissenschaftliche Wochenschift* 22, 673-679.
- Boekschoten, G. J. & Best, M. B. 1988. Fossil and recent shallow water corals from the Atlantic islands off western Africa. *Zoologische Mededelingen* 62, 99-112.
- Domning, D. P. 2005. Fossil Sirenia of the West Atlantic and Caribbean region. VII. Pleistocene *Trichechus manatus* Linnaeus, 1758. *Journal of Vertebrate Paleontology* 25, 685-701.
- Simpson, G. S. 1932. Fossil Sirenia of Florida and the evolution of the Sirenia. *American Museum of Natural History Bulletin* 59, 419-503.
- Shuker, K. P. N. 1995. The saga of the St. Helena sirenians. *Animals & Men* 4, 12-16.

Chapter 4:
Swan-necked seals

I am asked, quite frequently, about *Acrophoca longirostris*. What is it? It's a fossil seal from the Miocene and Pliocene Pacific coast of South America, described from good fossils by Christian de Muizon in 1981. Within the seal, or phocid, family, it belongs to a group called the monachines: the monk seals (monachins), elephant seals (miroungins), and Antarctic seals (lobodontins). As fossil tetrapods go, seals aren't regarded as particularly charismatic, nor do they attract much attention away from a handful of specialists. So what makes *Acrophoca* so interesting?

It's the contention that this animal should be regarded as a 'swan-necked seal': a pinniped that was sort of something like a mammalian version of a plesiosaur. Life restorations of *Acrophoca* are few are far between, but those that that have been published (on the cover of

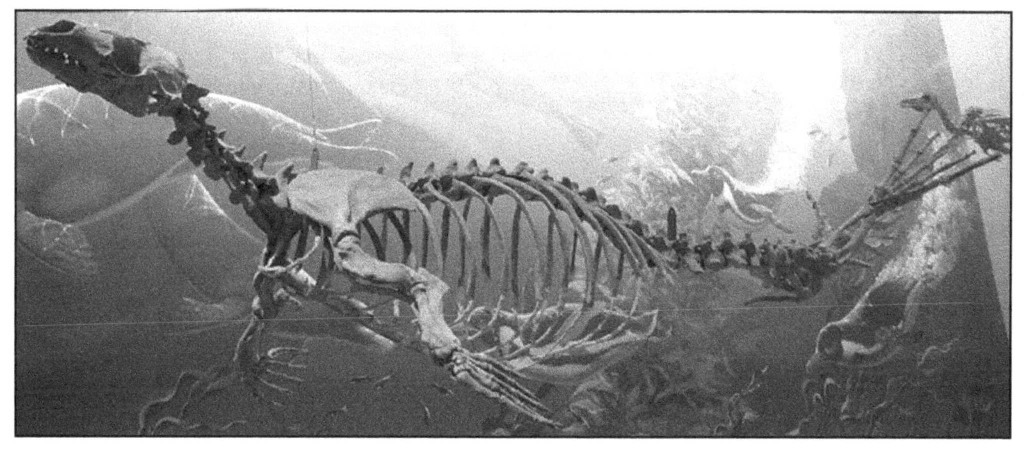

Fig. 12. A reconstructed skeleton of the fossil seal *Acrophoca longirostris*, as displayed at the Smithsonian National Museum of Natural History. Photo by Ryan Somma.

Muizon 1981, and in Naish 2001: reproduced in Fig. 13), or exist as museum murals (that at Karlsruhe's Staalische Museum für Naturkunde), do indeed make this seal look particularly long-necked compared to living phocids. Barnes *et al.* (1985) even went so far as to describe *Acrophoca* as 'serpentiform' (p. 41). All in all, it sounds like *Acrophoca* would have looked particularly odd. So here's the low-down.

In 2002, new *Acrophoca* material from Chile was described by Stig Walsh and myself. While, based on various detailed differences, it seemed likely that we had a species distinct from *A. longirostris*, we elected to leave the Chilean material as *Acrophoca* sp. This was sensible, because Stig's observations revealed that there was evidence for several *Acrophoca* species known from various horizons within the Miocene-Pliocene coastal sediments of Peru. Accordingly, we wrote "We have elected not to name [this] possible new species at this time because new Peruvian material of ... *Acrophoca* ... is under description and new species, represented by substantial remains, have been recovered (Muizon, pers. comm. 2000)" (p. 840).

Exactly this has turned out to be the case for some of the other marine mammals known from these sediments. The bizarre 'swimming sloth' *Thalassocnus*, for example, was originally described for a single species (*T. natans*) assumed to be a one-off. We now know that it was just one of several species within this genus, some of which were far more specialised for marine life than was *T. natans*. The youngest species, *T. carolomartini* and *T. yaucensis*, are poorly known, but what is known suggests that they were seal-like in some of their features, rather than sloth-like (Muizon *et al.* 2004). Given that one of the undescribed *Acrophoca* species is much longer-skulled than *A. longirostris* (Walsh & Naish 2002) – already notable for its long skull (hence the name) – it's conceivable that, as in *Thalassocnus*, some of the as-yet-unnamed species were altogether weirder looking. But we won't know this for sure until they get described.

Was *Acrophoca* really 'swan necked'? Well, judging from the specimens that Muizon described, it wouldn't have had a neck anywhere as near as long (proportionally) as that of a plesiosaur, but it still would have looked longer in the neck than any extant seal. Muizon's (1981) data shows that *Acrophoca* far exceeds other monachines in the length of its cervical vertebrae, and in total cervical column length. Based on similar-sized individuals, extant monachines have cervical column lengths of 218-249 mm, while *Acrophoca* is at

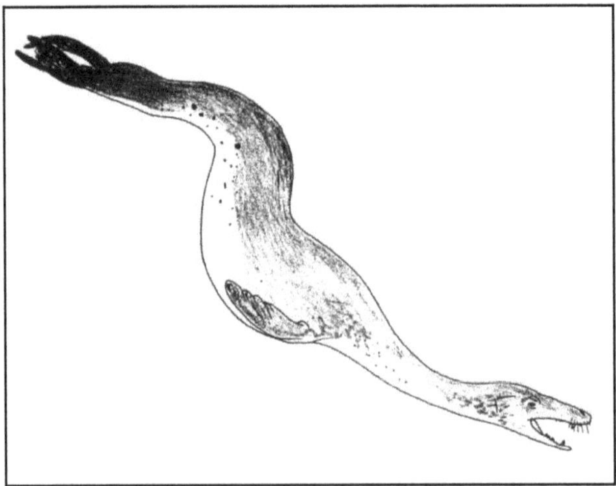

Fig. 13. *A. longirostris* as reconstructed by the author. This version probably makes the animal look too slender.

329 mm. King (1983) showed that in most seals, cervical column length is between 17 and 19% of the total length of the vertebral column – not different from terrestrial carnivorans like dogs (also 17%). In Muizon's skeletal reconstruction, the neck length of *Acrophoca* is about 21% of total vertebral column length. So, yes, a bit longer-necked than living seals, but 'swan-necked'? Well, even living pinnipeds that we don't regard as particularly long-necked have necks flexible enough to allow a startling lengthening effect when they lunge, stretch or spy-hop. Bonner (1994) noted that a seal can "extend its neck suddenly when it strikes prey. Seals can do the same thing on land, as many a seal researcher has discovered to his or her cost" (p. 18). This applies particularly for some otariids and for lobodontins like *Hydrurga*. What I'm getting at is that, when alive, *Acrophoca* would have been capable of looking even longer in the neck than we might think just from its fossils. But clearly it's a stretch to imagine this animal as having a long long long neck like a swan, or a plesiosaur, so, sadly, 'swan-necked seal' really is a bit of an exaggeration.

Fig. 14. Another life reconstruction of *Acrophoca longirostris* (by Arthur Weasley), this time giving it a more robust, *Hydrurga*-like appearance.

The phylogenetic affinities of *Acrophoca* might also tell us something about its appearance in life. The problem is that this is a particularly contentious topic among seal experts. Muizon (1981) regarded *Acrophoca* as a lobodontin: that is, a member of the group that includes *Hydrurga* (the leopard seal), *Ommatophoca* (Ross's seal) and *Lobodon* (the crabeater seal), and specifically as the sister-taxon to *Hydrurga*. Though it's been suggested that lobodontins might be paraphyletic with respect to *Monachus* (a highly problematic genus within seal phylogeny), most recent studies have supported lobodontin monophyly (Bininda-Emonds *et al.* 1999). If *Acrophoca* is a member of this group, as Muizon (and Walsh & Naish) concluded, then we should be thinking of living lobodontins, and specifically *Hydrurga*, as the closet extant models. And I find that particularly interesting, because it's my opinion that *Hydrurga* is one of the most 'disturbing' looking of all mammals. It's a chunky mother of a seal: thickset and quasi-reptilian, with a theropod-like jaw line and robust skull demarcated from its neck. I

would imagine *Acrophoca* as a paradoxical combination of these traits: quasi-reptilian and *Hydrurga*-like on one hand, but also longer-skulled, longer-necked and overall more gracile. It would have been odd, and don't forget that other species in the genus would have looked even odder.

Refs - -

- Barnes, L. G., Domning, D. & Ray, C. E. 1985. Status of studies on fossil marine mammals. *Marine Mammal Science* 1, 15-53.
- Bininda-Emonds, O. R. P., Gittleman, J. L. & Purvis, A. 1999. Building large trees by combining phylogenetic information: a complete phylogeny of the extant Carnivora (Mammalia). *Biological Reviews* 74, 143-175.
- Bonner, N. 1994. *Seals and Sea Lions of the World.* Blandford (London), pp. 224
- King, J. E. 1983. *Seals of the World.* British Museum (Natural History) (London), pp. 240.
- Muizon, C. de 1981. Les vertébrés fossiles de la formation Pisco (Pérou). Première partie: deux nouveaux Monachinae (Phocidae, Mammalia) du Pliocene de Sud-Sacaco. *Travaux de l'Insitut Français d'Études Andines* 22, 1-161.
- Muizon, C. de, McDonald, H. G., Salas, R. & Urbina, M. 2004. The youngest species of the aquatic sloth *Thalassocnus* and a reassessment of the relationships of the nothrothere sloths (Mammalia: Xenarthra). *Journal of Vertebrate Paleontology* 24, 387-397.
- Naish, D. 2001. Sea serpents, seals and coelacanths: an attempt at a holistic approach to the identity of large aquatic cryptids. In Simmons, I. & Quin, M. (eds) *Fortean Studies Volume 7*. John Brown Publishing (London), pp. 75-94.
- Walsh, S. A. & Naish, D. 2002. Fossil seals from late Neogene deposits in South America: a new pinniped (Carnivora, Mammalia) assemblage from Chile. *Palaeontology* 45, 821-842.

Chapter 5:
The bear-eating pythons of Borneo

"After being poked with a stick, it fled into a nearby stream, producing, as it twisted, the sound of breaking bones" (Fredriksson 2005, p. 166). While I don't want people to think that I'm strangely obsessed with the macabre details of unusual predator-prey relationships (see Chapter 1), I will admit that it's lines like that that prove especially rewarding when reading behavioural studies on predation. This excerpt is from Gabriella Fredriksson's remarkable recent paper on the predation of Sun bears *Helarctos malayanus* by big Reticulated pythons *Python reticulatus*. Yet again, I am amazed by the habits of a big tetrapod predator... I mean: pythons killing bears?

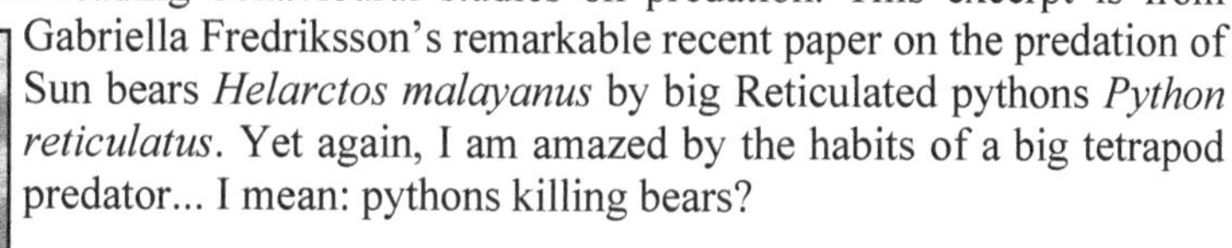

Fig. 15. The Sun bear is a small bear that occurs throughout the tropical forests of southeast Asia. It has a particularly long tongue. This captive individual was photographed at the Columbus Zoo, Powell, Ohio, by Ryan E. Poplin.

Granted, *Helarctos* is the smallest living bear, with Nowak giving total length as c. 1-1.5 m, shoulder height as 70 cm, and weight as 27-65 kg (Fig. 15). Though it's been tradition in recent decades to sink all bears into *Ursus*, *Helarctos* is used here as "the monophyly of *Ursus* is not assured, which supports continuing calls for a revision of the genus-level taxonomy of ursids" (Bininda-Emonds *et al.* 1999).

Fredriksson is a Netherlands-based *Helarctos* specialist who has previously published reports on their conservation and on their interactions with humans. As part of her research she's been looking after confiscated cubs, later releasing them into suitable habitat.

These bears wear radiocollars. Two attacks by retics on *Helarctos* are reported in the paper. One is an assumed attack: a 31-kg bear was sleeping in a tree, called out in distress during the night (*Helarctos* is diurnal), and was later observed with wounds suggestive of a snake attack. Other possible attackers were eliminated from consideration, though the presumed snake itself was not observed.

The second case is, well, just a little more convincing. In July 1999 a second radiocollared bear (weight 23 kg) remained motionless for about 4 hrs, a period of time suggestive either of a dropped radiocollar, or of death. Fredriksson tracked the collar and found "that the signal was being emitted from the stomach of a large python". Because the collar still functioned, Fredriksson and her colleagues and helpers could now track the snake, though to be honest it didn't do much over the next few days. They ended up catching it and holding it in captivity in the hope that the collar would be regurgitated. The snake only escaped once (and had to be re-tracked). By late October the snake still had the collar inside it, so the decision was made to remove it surgically. All of this went to plan and the snake was fine: it was released in late November. I was relieved that at no point in the paper was the term 'dispatched' or 'euthanized' used... granted, all field biologists will tell you that it is not their business to interfere with acts of predation carried out upon their study subjects, but, in practice, things very occasionally work out differently.

Fig. 16. A Reticulated python.

At a very impressive 6.95 m and 59 kg, this snake was about three times bigger than the bear it ate, and thus quite conceivably capable of what it did. Furthermore, the bear was in poor condition, being underweight due to a local fruit shortage and suffering stress from having nursed a small cub (and, sad to say, the cub was nowhere to be found after the adult bear's death). Having said all that, killing and eating a bear – big and sharp curved claws, strong jaws and all that – still isn't going to be a picnic. It probably doesn't happen often, though maybe it happens more often than we would have thought. Fredriksson cites one other reported instance, also on Borneo.

Though this is the first technical report on bear predation by pythons, other macro-mammal predation acts have been long known. Isemonger (1962) reported cases where African pythons killed and swallowed antelopes that must have weighed around 20 kg, a 32-yr-old man was allegedly eaten by a 7-m retic in 1998 (Auliya unpublished, cited in Fredriksson 2005), and Shine *et al.* (1998) reported a case of a retic swallowing a pig (of unstated species) that weighed 60 kg. If the latter case is valid (it was only reported to Shine *et al.* as a pers. comm. from a Sumatran snake-skinner, and is thus an unverified anecdote), then it would be a new world record, as the largest verified mammal meal according to Carwardine's *Guinness Book of Animals Records* was a 59 kg impala eaten by an African rock python *Python sebae.*

But it seems that some big pythons are sometimes over-ambitious, attempting to swallow animals that are too big even for their distensible jaws. So, in the 1995 Malaysian case of a 6.65 m python that tried to swallow 29-yr-old Ee Heng Chun, the snake failed to complete the swallowing. This snake reportedly weighed 140 kg, and after being scared away from the body was shot dead by police. Other snakes may consume animals far too large for their own digestive abilities, and they then become immobilised and stricken by heatstroke (Klauber 1982), or may swallow another snake powerful enough to administer a dangerous bite to the swallower's stomach wall.

Fig. 17. Record-holding Reticulated pythons can supposedly approach 10 m in length, though any length over 8 m is truly exceptional.

Other pythons have made the mistake of swallowing horned antelopes, with the result being that the antelope's horns have fatally pierced the stomach and body wall (Mattison 1995). Such piercings are not always fatal: remarkably, the injuries may heal after the offending horns drop off as the prey's body decomposes inside the snake (Isemonger 1962). 'Death by piercing' isn't limited to big snakes: Klauber (1982) reported that horned toads *Phrynosoma* may kill snakes that try to swallow them (the lizard's horns perforate the snake's throat) and Ramírez-Bautista & Uribe (1992) described the case of a Lyre snake *Trimorphodon biscutatus* (a terrestrial colubrid) that died after the spiny tail scales of a Spiny-tailed iguana *Ctenosaura pectinata* pierced the snake's stomach and oesophagus.

So... snakes eating bears, snakes eating – and trying to eat – people, and snakes getting pierced by horns... don't say you're not getting your money's worth.

Refs - -

- Bininda-Emonds, O. R. P., Gittleman, J. L. & Purvis, A. 1999. Building large trees by combining phylogenetic information: a complete phylogeny of the extant Carnivora (Mammalia). *Biological Reviews* 74, 143-175.
- Fredriksson, G. M. 2005. Predation on sun bears by reticulated python in east Kalimantan, Indonesian Borneo. *The Raffles Bulletin of Zoology* 53, 165-168.
- Isemonger, R. M. 1962. *Snakes of Africa: Southern, Central and East.* Thomas Nelson and Sons (Africa), Johannesburg.

- Klauber, L. M. 1982. *Rattlesnakes: Their Habits, Life Histories, and Influence on Mankind.* University of California Press, Berkeley, Los Angeles, and London.
- Mattison, C. 1995. *The Encyclopedia of Snakes*. Blandford, London.
- Ramírez-Bautista, A. & Uribe, Z. 1992. *Trimorphodon biscutatus* (Lyre snake): predation fatality. *Herpetological Review* 23, 82.
- Shine, R., Harlow, P. S., Keogh, J. S. & Boeadi. 1998. The influence of sex and body size on food habits of a giant tropical snake, *Python reticulatus*. *Functional Ecology* 12, 248-258.

Chapter 6:

They bite, they grow to huge sizes, they locate human corpses: the snapping turtles

Until recently I had very little to do with turtles. I'd written some brief pieces on the taxonomy of Indian Ocean giant tortoises, and naturally I've had to do consultancy work on giant fossil taxa, like the Cretaceous protostegids and *Stupendemys*, a Venezuelan pleurodire that exceeded 3 m in length and must have weighed a couple of tons. But it's the sharing of my office with friend and colleague Sarah Fielding that has brought me into the surreal Gaffneyesque world of turtles more than before, and, oh, the adventures I've had…

We have the fascinating tale of Branston the pickled turtle and the sorry trade in African helmeted turtles *Pelomedusa subrufa*, my adventures with Sarah, Dave and a new araripemydid likened to Jennifer Lopez (Fielding *et al.* 2005), and, best of all, the story of the perplexing, colour-changing Cuthbert, an as-yet-unidentified emydid who lived in my office before being driven specially down to Exeter. He died, and is now stuck in a freezer. A deadline for a short manuscript dealing with the pleurodires of the Brazilian Crato Formation has come and gone, but there isn't really a desperate rush on it and I'll be ok. I've written most of it.

Fig. 18. American snapping turtle. Aww, cute!

Fig. 19: Alligator snapping turtle. The head is proportionally larger than that of *Chelydra* and decorated with small papillae, and the carapace is more rugose and strongly keeled.

Among the most interesting turtles of them all are the snappers: the two extant members of Chelydridae, *Chelydra serpentina* (the American snapping turtle or Common snapping turtle) and *Macroclemys temminckii* (the Alligator snapping turtle) [incidentally, the Big-headed turtle *Platysternon* has been included at times within this family, though whether it's even close to chelydrids is now disputed]. Bizarre skulking benthic aquatic ambush predators, they are famous for being cryptic, for pretending to look like the lumpy bumpy muddy substrate they camouflage themselves on, and for biting pretty much most things that come within reach. Alligator snappers apparently only leave the water to bask or lay eggs, while snappers will walk for long distances across land (up to 16 km) to find suitable breeding sites.

Snappers (from hereon used specifically for *Chelydra*) are omnivorous, and as well as all manner of invertebrates and small vertebrates, will also eat various plants. They get to about 50 cm in carapace length and have three low keels running the length of the carapace. This strikes me as a curious parallel with leatherbacks, and also with some placochelyid placodonts, but I don't think this means much as leatherbacks are pelagic deep-divers that swim incessantly and prey on cnidarians, while placochelyids appear to have been durophagous benthic foragers that didn't spend their time sitting on the substrate. Anyway, I digress.

With a range extending from Nova Scotia and southern Quebec to Ecuador, snappers are reasonably variable, and populations differ in the proportional size of various of their scutes, and how their necks are ornamented. Again we come across what is becoming a frustratingly familiar theme: are they just one species? This has proved to be quite a contentious area. Traditionally, the species has been divided into four supposed subspecies: *C. serpentina serpentina* (of continental Canada and the US), *C. s. acutirostris* (of Central America and northern South America), *C. s. rossignonii* (of Mexico, Belize and nearby) and *C. s. osceola* (restricted to Florida). Feuer (1971) argued that *C. s. serpentina* and *C. s. osceola* graded into one another, and that the latter was therefore not worthy of distinction. Other workers have expressed doubts about the supposed distinction of the other subspecies. However, Phillips *et al.* (1996) found that *C. s. serpentina* and *C. s. osceola* were distinct taxa, and more closely related to each other than to *C. s. acutirostris* and *C. s. rossignonii*. Furthermore, the mtDNA data they compiled suggested that both these latter forms should be regarded as distinct species. What was once the one species *C. serpentina* therefore becomes three, if Phillips *et al.* (1996) are right.

Are they right? Well, their conclusions were seriously challenged by Sites & Crandall (1997) who argued that Phillips *et al.* (1996) did not "present any species concept as a testable hypothesis" and that "data [was] collected in the absence of any conceptual framework for diagnosing species boundaries" (p. 1289). Sites & Crandall (1997) didn't support species status for the 'taxa' recognised by Phillips *et al.* (1996), though they did suggest that snappers from Ecuador might deserve specific recognition. More recently, I note that Walker *et al.* (1998) have published a paper with the title 'Phylogeographic uniformity in mitochondrial DNA of the snapping turtle': I haven't seen this but it at least implies that they did not find evidence for polytypy within snappers. So right now things are undecided, and *Chelydra* may or may not turn out to be polytypic.

Alligator snappers are altogether different from snappers. On the taxonomic side of things, check the literature and you see an annoying amount of inconsistency on the generic name, with authors switching between *Macroclemys*, *Macroclemmys* and *Macrochelys*. I'll assume Ernst & Barbour (1989) are right – they usually are – and that the two variants arose as typos. They (alligator snappers, not Ernst & Barbour) have a carapace with very tall keels, each of which possesses high, sub-pyramidal knobs, a far more massive and nasty-looking head, and overall they're more gnarly looking, with papillae and skin fronds projecting from their limbs, neck and elsewhere. Unlike snappers, they sport a worm-like lure on the tongue. By sitting with the mouth open, and by waggling the lure, they... err... lure in prospective prey.

What's more, this is one of the biggest freshwater turtles in the world, reaching 90 cm and 100 kg at least. Conant & Collins (1991) give a maximum verified weight of 143 kg. Other freshwater turtles are longer, but not as heavy (though read on). Asian giant softshells *Pelochelys bibroni*, for example, may exceed 120 cm. Ok, so as it happens there are the recently reported enormous softshells of Hoan Kiem Lake in Hanoi, Vietnam, which, according to published photos, appear to be about the size of a car (well, nearly. They're said to be up to 2 m long). They're clearly the world's biggest freshwater turtles, but right now their size isn't verified, nor do we even know what they are (they're probably *Rafetus swinhoei*, but it's also been proposed that they represent a new species, dubbed *R. leloii*).

Getting back to alligator snappers, there are, as you'd expect, lots of anecdotal tales of animals bigger than the 'verified' ones, and while none of these rumoured individuals have been confirmed, they're fun to think about. In 1937 a 183 kg specimen was supposedly caught in the Neosho River in Kansas. More famous – it might be the most famous alligator snapper of all time, though whether or not it was a real animal is open to question – was the gigantic individual, dubbed variously the 'Beast of Busco' or 'Oscar' (after Oscar Fulk, allegedly the first witness), that supposedly dwelt in Fulk's Lake, Churubusco, Indiana. 'It' (supposedly the same individual) was seen on numerous occasions since 1898. A 1949 sighting by farmer Gale Harris described it as about 1.8 m long and 1.5 m wide, and numerous attempts to catch it ensued. These involved the building of a stockade around the turtle, attempted capture by divers, the use of a dredging crane and pumps, and a female sea turtle imported to act as bait (Newton 2005). Apparently two people died in one of these dumb schemes. Needless to say the animal was never measured. It was also claimed to be 'the size of a dining-room table' and estimated to be nearly 230 kg.

Well, ok, maybe.

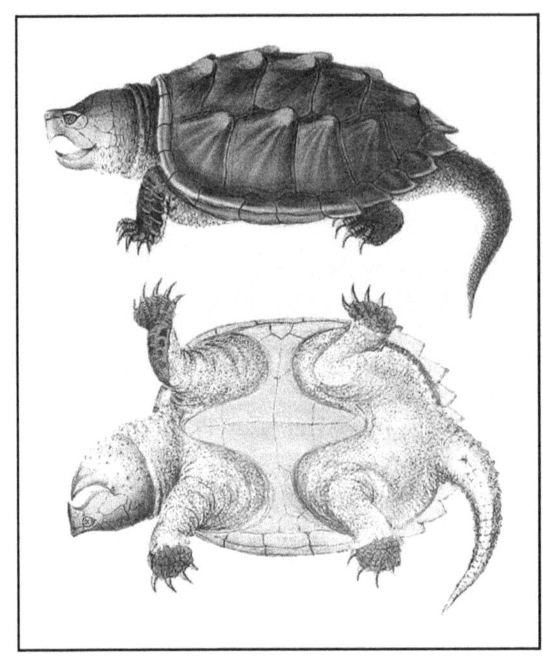

Fig. 20. This old illustration of an Alligator snapper dates to 1842 and was produced by J. H. Richard del for John Holbrook's 1842 volume *North American Herpetology; or, A description of the Reptiles Inhabiting the United States.*

Hyperexcitability, supercooling, and the inevitable recolonisation of Europe in the Anthropocene

It turns out that there are questions, not just about how big alligator snappers get, but also how far north they might occur. While, 'officially', they don't get further north than Iowa, there are reports of alligator snappers from Canada (Anon. 2003). You might think that these were snappers (as these are definitely present that far north), but witnesses say that the animals were just too big for that (you'd also think they'd report the distinctive eggbox-like carapace knobs, but no details on that). The northerly occurrence of snappers is interesting, as not only are they cold-tolerant, they're *really* cold tolerant, and able to tolerate literally freezing conditions. They hibernate under water (Ultsch 1989), but they'll also remain active at temperatures as low as -2 to -4°C. They can tolerate supercooling down to -6°C, and specimens cooled to as low as -9°C have remained alive (Costanzo *et al.* 1999).

Clearly, they are adapted for dealing with low temperatures. Like other cold-adapted reptiles and other ectothermic vertebrates, they exhibit behavioural hyperexcitability when chilled (that is, they suddenly become apparently alert and vigorous), and (counter-intuitively) increase the number of movements they make when exposed to really low temperatures. Apparently, hyperexcitability and an increase in rate of movement 'may stem from failure of inhibitory synapses, which are more thermally labile than excitatory ones' (Costanzo *et al.* 1999, p. 165; Prosser & Nelson 1981). This may be advantageous as it encourages the animal to move away from a potentially lethal environment. To restate that in more simple terms: extreme cold makes the animals become twitchy and restless, and this perhaps encourages them to move elsewhere. This behaviour is thought to be adaptive as it doesn't occur in species than don't have to deal with extreme cold, nor does it occur in species that are even more cold tolerant than snappers (e.g., Painted turtles *Chrysemys picta*).

So this brings us on to something else: given that they're kept as pets a lot in Europe, and given that – like all pet turtles - they're good at getting themselves released (i.e., there's something about turtles that makes people want to go chucking them out into the wild), will we end up having them as ferals in Britain and elsewhere? The answer is yes, and... they're already here, at least judging by the number of individuals that get reported (Beebee & Griffiths 2000).

Already there is much concern over the many feral Red-eared sliders *Trachemys scripta* we have in Britain. Though it might seem 'nice' to have sliders in the wild, they're presumably eating lots of our already beleaguered fauna, and preliminary studies indicate that they have a heavy impact on lissamphibians. The ecological impact of snappers – bigger, nastier, hungrier – would be greater. At the Zoologica 1997 Convention I spoke to a group of ecologists who were studying the impact that feral freshwater turtles are having on British natives. I took the free paperwork they provided, but have since lost it. They confirmed that there are now indeed enough feral snappers for us to start worrying, though I don't think they had an accurate head count or anything. But you can verify that they're there by just keeping an eye on the press: in July 2003 a snapper was captured near Walsall in the West Midlands, for example.

But if they are here, could they maintain populations, or breed? Alderton (1988) thought that this was unlikely, stating that the climate here is too cool to allow enough successful breeding to occur. However, snappers can maintain viable populations "where the growing season has at least 100 frost-free days per year" (Tarduno *et al.* 1998). Most of Britain has less than 100 frost-free days per year, so that criterion is fulfilled. More generally, Tarduno *et al.* (1998) wrote that viable populations "do not occur in areas with a warm-month average maximum temperature of less than 25° C. This measure corresponds to a warm-month mean temperature of 17.5° C and a mean annual temperature of 2.5° C" (p. 2241). After some internet research on the British climate, I found that England has a warm-month average maximum temperature of about 26° C, a warm-month mean temperature of 15.6° C, and a mean annual temperature of between 8.5 and 11° C. The mean warm-month temperature is thus a little too low for snappers, but only a bit too low. It's therefore dangerously close to likely that snappers could survive here forever, without trouble (well, until the next glacial cycle kicks in, or until the North Atlantic gyre switches off), and as our mean temperature is set to rise, this is going to be even more true. Add to this the fact that snappers live for decades: say there are feral individuals here now, able to survive fine but not to successfully breed... well, given that they'll still be here in a few decades time, they have the potential to stock our future waterways with their hatchlings.

This brings us on to something else. Though restricted to the Americas today, chelydrids once ruled the world, with numerous fossil species inhabiting Europe, Asia and possibly Africa (Gaffney & Schleich 1994). Why did they become extinct in these areas? To my knowledge, such questions have only been asked for the European fauna. Here, chely-

Fig. 21. Snapping turtles once ruled the world. Another Common snapping turtle, this time crossing a road.

drids are present right up to the middle Pliocene (Broin 1977) at least… in fact, Delfino *et al.* (2003) put a question mark for the presence of the group (their fig. 1) in the late Pliocene, implying that there might be late Pliocene European fossils of the group, and Gaffney & Schleich (1994) stated that *Chelydropsis* extended into the Pleistocene. In asking "what happened to the herpetofauna?", Delfino *et al.* (2003) suggested that the loss of chelydrids (and other taxa, including cryptobranchid salamanders, varanids and elapids) from Pliocene Europe might have been triggered by the climatic cooling that kicked in at this time. While living snappers are cold tolerant, they cannot persist where mean annual temperatures are too low, as we saw above.

So glacial conditions appear to have caused their extinction in Europe, and presumably in Asia too. Why didn't they persist in southern Asia, the Middle East, or Africa, where they also occur as fossils? Good question.

Bite, lunge, lure and snap

What perhaps makes chelydrids most interesting is what they eat, or at least what they bite. "Large individuals are known to have caused injury to people unwary enough to step into or swim in the water near them, and they are quite capable of removing a toe or finger if given the opportunity" (Alderton 1988, p. 112). Indeed, they are anecdotally credited with being able to bite through a broom handle. For his excellent TV series *O'Shea's Dangerous Reptiles*, Mark O'Shea decided to test this dubious assertion. With the help of a colleague he caught the largest and nastiest alligator snappers he could, pissed them off by poking them, and then pissed them off some more by shoving a broom handle into their mouths. All the turtles bit happily, and bit hard. And as impressive as it was, sad to say not one turtle was powerful enough to cleave neatly through 25 thick mm of solid wood, which to be honest isn't much of a surprise.

Rather more rigorous tests were applied by Herrel *et al.* (2002) who tested the bite force of numerous diverse turtles, including both snappers and alligator snappers. They found snappers to have bite forces of between 208 and 226 Newtons, and alligator snappers of between 158 and 176 N.

This powerful bite can have its practical uses. Zoo vet David Taylor, one of the best sources for zoological anecdotes (think tapir-eating hippos and a zebra with a tooth embedded in one of its testicles), tells the story in one of his books of how an alligator snapper in Indiana was used by its Native American owner to retrieve lost human corpses. The turtle (which was kept on a long leash) was taken to the lake where the person was missing, and released. Eventually its owner would know from the pull on the leash that the turtle was tugging around on a carcass, somewhere on the bottom. It would then be pulled slowly to the surface, and it would successfully bring the corpse with it. I couldn't find the relevant Taylor book that relates this tale, but I did find a version of the story in Alderton (1988).

Ordinarily snapping turtles don't feed on broom handles or human corpses. They are sit-and-wait predators that lunge with great speed at passing objects, but they also consume static ob-

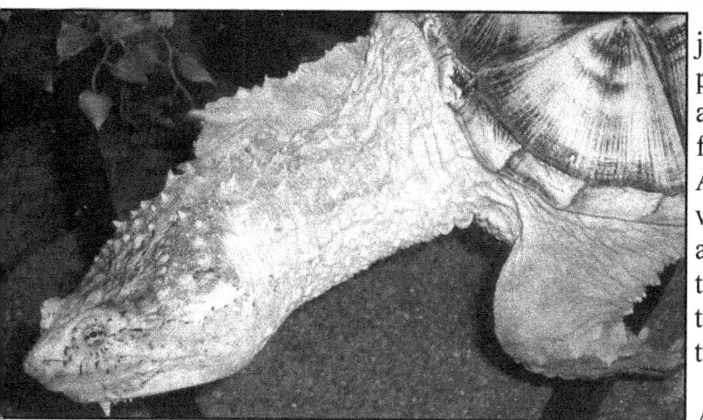

Fig. 22. A close-up view of an unusually clean, and pale, captive Common snapper.

jects like molluscs and some plants. *Chelydra* likes to skulk in aquatic vegetation, though it also floats near the water surface. Aquatic insects, crustaceans, worms and fish are eaten, and it also preys on a diversity of tetrapods. These include frogs and toads, salamanders, snakes, smaller turtles, birds and small mammals.

As mentioned, they also eat various plants, including Canadian pondweed and parts of water lilies, and they are also reported to eat snails, clams and freshwater sponges. This interests me, as if they eat plants and clams and sponges then they obviously don't restrict their feeding activities to moving objects. Do they use olfaction to determine whether these non-moving objects are edible? I don't know, and I can't find an answer in the literature. They are reported to consume more plant material during the warmer months, when plant growth is more abundant.

Macroclemys, the alligator snapper, is altogether different from *Chelydra* in terms of what it does and what it can do. As is reasonably well known, it possesses a lure-like organ on the floor of its mouth. This isn't the tongue as sometimes stated but a specialised mobile organ attached to the tongue at mid-length. Grey at rest, it is red when active, and the turtles deploy it by sitting still with the mouth wide open, and by wriggling it. Some herpetologists have proposed that the lure is only used during the daytime when the turtles sit passively on the substrate (as opposed to the nightime, when the turtles actively forage for prey and don't, in theory, deploy the lure), and while there are some observations recorded from captive individuals, it doesn't seem that much is known about the use of the lure.

Like *Chelydra*, *Macroclemys* grabs crustaceans, worms, fish, frogs and snakes, and it also eats static prey like plants and clams. It is more dedicated to turtle-killing than *Chelydra* is, and well able to catch and kill turtles that are about as big as it is. Allen & Neill (1950) reported alligator snappers to eat Painted turtles *Chrysemys picta*, Chicken turtles *Deirochelys reticularia*, mud turtles *Kinosternon*, as well as other alligator snappers. In fact, their predation on smaller turtles may be significant enough to keep populations of other species low in areas where alligator snappers are abundant. As suggested by that story about the corpse-finder alligator snapper, members of this species are also adept at carrion feeding, though how important this is in their natural diet is uncertain.

Unless you already know the answer (clever you), you might be wondering how slow, clumsy turtles are able to catch speedy, agile aquatic prey like fish. The feeding styles of aquatic turtles are pretty fascinating, and have been studied recently by Patrick Lemell and colleagues (see Lemell *et al.* 2000, 2002). Using expandable throats, rapidly opening jaws and stream-

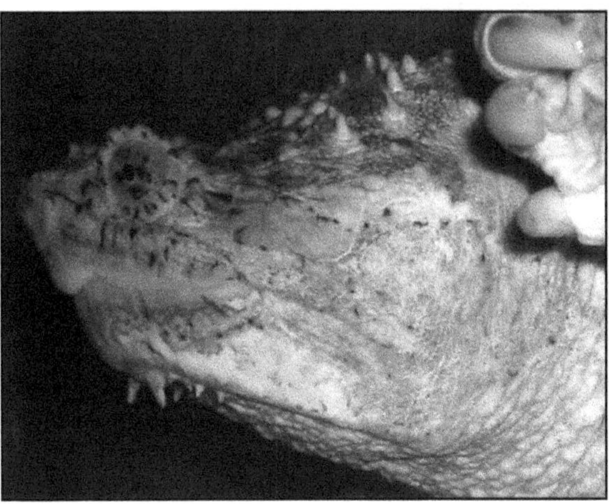

Fig. 23. Another close-up of the same Common snapper shown in Fig. 22. Note the barbel-like growths on the chin and soft 'horns' at the back of the head.

lined skulls, aquatic turtles can dart their head forward, open their jaws, and engulf water and prey within – literally – a fraction of a second. Along these lines, Lauder & Prendergast (1992) used high-speed video recording to study feeding behaviour in *Chelydra*. They showed that *Chelydra* lunges and engulfs fish prey within 78 ms, with "peak head extension velocities of 152.5 cm per second". It doesn't lunge as quickly when feeding on worms, darting its head forward at a sedate 54 cm per second in these cases, and engulfing worms within a leisurely 98 ms. In contrast to matamatas and other aquatic turtles that generate massive negative pressures and thereby employ suction to engulf prey, *Chelydra* is predominantly a ram-feeder that doesn't generate negative pressure when it lunges.

I'd like to talk more about aquatic feeding behaviour in turtles, but it'll have to wait for another time. It's a story that involves plethodontid salamanders, placodonts and the evolution of filter-feeding.

Refs - -

- Alderton, D. 1988. *Turtles & Tortoises of the World*. Blandford, London.
- Allen, E. R. & Neill, W. T. 1950. The alligator snapping turtle, *Macrochelys temminckii*, in Florida. *Special Publication of Ross Allen's Reptile Institute* 4, 1-15.
- Anon. 2003. Turtl-ey amazing. *Animals & Men* 31, 12-13.
- Beebee, T. & Griffiths, R. 2000. *Amphibians and Reptiles*. HarperCollins, London.
- Broin, F. De 1977. Contribution à l'étude des chéloniens. Chéloniens continentaux du Crétacé et du Tertiaire de France. *Mémoires du Muséum national d'Histoire Naturelle* C 38, 1-366.
- Conant, R. & Collins, J. T. 1991. *A Field Guide to Reptiles and Amphibians of Eastern and Central North America (Third Edition)*. Houghton Mifflin (Boston), pp. 450.
- Costanzo, J. P., Litzgus, J. D. & R. E. Lee. 1999. Behavioral responses of hatchling painted turtles (*Chrysemys picta*) and snapping turtles (*Chelydra serpentina*) at subzero temperatures. *Journal of Thermal Biology* 24, 161-166.
- Delfino, M., Rage, J. C. & Rook, L. 2003. Tertiary mammal turnover phenomena: what happened to the herpetofauna? *Deinsea* 10, 153-161.
- Ernst, C. H. & Barbour, R. W. 1989. *Turtles of the World*. Smithsonian Institution

Press, Washington, D. C. & London.

- Feuer, R. C. 1971. Intergradation of the snapping turtles *Chelydra serpentina serpentina* (Linnaeus, 1758) and *Chelydra serpentina osceola* Stejneger, 1918. *Herpetologica* 27, 379–384.
- Fielding, S., Martill, D. M. & Naish, D. 2005. Solnhofen-style soft-tissue preservation in a new species of turtle from the Crato Formation (Early Cretaceous, Aptian) of north-east Brazil. *Palaeontology* 48, 1301-1310.
- Gaffney, E. & Schleich, H. H. 1994. New reptile material from the German Tertiary. 16. On *Chelydropsis murchisoni* (Bell, 1892) from the Middle Miocene locality of Unterwohlbach/South Germany. *Courier Forschungsinstitut Senckenberg* 173, 197-213.
- Herrel, A., O'Reilly, J. C. & Richmond, A. M. 2002. Evolution of bite performance in turtles. *Journal of Evolutionary Biology* 15, 1083-1094.
- Lauder, G. V. & Prendergast, T. 1992. Kinematics of aquatic prey capture in the snapping turtle *Chelydra serpentina. Journal of Experimental Biology* 164, 55-78.
- Lemell, P., Beisser, C. J. & Weisgram, J. 2000. Morphology and function of the feeding apparatus of *Pelusios castaneus* (Chelonia; Pleurodira). *Journal of Morphology* 244, 127-135.
- Lemell, P., Lemell, C., Snelderwaard, P., Gumpenberger, M., Wocheslánder, R. & Weisgram, J. 2002. Feeding patterns of *Chelus fimbriatus* (Pleurodira: Chelidae). *The Journal of Evolutionary Biology* 205, 1495-1506.
- Newton, M. 2005. *Encyclopedia of Cryptozoology: A Global Guide to Hidden Animals and Their Pursuers*. McFarland & Company, Jefferson, North Carolina.
- Phillips, C. A., Dimmick, W. W. & Carr, J. L. 1996. Conservation genetics of the Common snapping turtle (*Chelydra serpentina*). *Conservation Biology* 10, 397-405.
- Prosser, C. L. & Nelson, D. O. 1981. The role of nervous systems in temperature adaptation of poikilotherms. *Annual Review of Physiology* 43, 281-300.
- Sites, J. W., Jr., & Crandall, K. A. 1997. Testing species boundaries in biodiversity studies. *Conservation Biology* 11, 1289-1297.
- Tarduno, J. A., Brinkman, D. B., Renne, P. R., Cottrell, R. D., Scher, H. & Castillo, P. 1998. Evidence for extreme climatic warmth from Late Cretaceous Arctic vertebrates. *Science* 282, 2241-2244.
- Ultsch, G. R. 1989. Ecology and physiology of hibernation and overwintering among freshwater fishes, turtles, and snakes. *Biological Reviews* 4, 435-516.
- Walker, D., Moler, P. E., Buhlmann, K. A. & Avise, J. C. 1998. Phylogeographic uniformity in mitochondrial DNA of the snapping turtle (*Chelydra serpentina*). *Animal Conservation* 1, 55–60.

Chapter 7:
Tortoises that drink with their noses, or: alas, goodbye Hololissa?

My renewed recent interest in the members of Testudines (and, oh no, you can't call them Chelonia anymore) led me, in between writing about obscure English sauropods, to dig around on a hard drive for the following text on giant tortoises from the Indian Ocean. It's a bit dated, having first appeared on a mailing list a few years ago, but I've added a bit to it and tried to incorporate new information (since it was written Paul Chambers's book on giant tortoises has been published, but I haven't read that yet – it's called *A Sheltered Life* – and Justin Gerlach's 2004 *Giant Tortoises of the Indian Ocean* has appeared). The main topic here is the systematics of the giant tortoises of the western Indian Ocean, specifically those of Madagascar, the Comoros, Aldabra and the Seychelles. As yet I haven't seen much made of this information in print, nor on the internet.

Fig. 24. A Seychelles giant tortoise of the sort identified by some as *Dipsochelys hololissa*. Photo by Thierry Caro.

Recent studies, discoveries and taxonomic revisions have meant that the number of Indian Ocean tortoise species, and the genera into which these species are placed, has increased somewhat. Those previously lumped together as *Geochelone gigantea* are today placed in the genus *Dipsochelys* Bour, 1982 (aka *Aldabrachelys*

Loveridge & Williams, 1957, which has been argued by some to be the correct generic name for these animals, despite Bour's contention that it is unavailable). While this idea is widely accepted, exactly how many species belong within *Dipsochelys* is more controversial, as we'll see. Morphologically, these are very interesting tortoises, with a particularly short supraoccipital crest and a notably deep, narrow snout with unusual nasal passages. Normally in tortoises, the channel connecting the external naris to the olfactory chamber is short and sub-horizontal, and the olfactory chamber is also short. In *Dipsochelys*, however, the external naris is vertically elongated and the nasal passage is long and ascends steeply as it approaches the olfactory chamber. The chamber itself is then quite long, and demarcated from the nasal passage by a distinct medially projecting vertical ridge. What is this all about?

It seems that these are specialisations allowing *Dipsochelys* to drink… through its nostrils. As reported by Nick Arnold (1979), I. R. Swingland found that *Dipsochelys* sometimes drinks through its nose, and that these unusual features assist it in doing this. The narrow, pointed snout is better for getting the nostrils into small puddles and other bodies, and the vertical ridge at the front of the olfactory chamber seems to anchor a soft-tissue valve that prevents water from being snarfed deep into the olfactory chamber (as would happen with any normal tortoise, were it to try drinking like this). This all makes sense in a group of tortoises that have to make a living in arid environments where standing water is relatively rare. Incidentally, since this was written, Gerlach (2004a) has published the paper 'The complex nasal structure of *Dipsochelys* and its identification as a true Jacobson's organ'.

But while, as noted above, most workers – until recently – had regarded these tortoises as representing a single species, things have now become much more complicated. In 1982, Roger Bour proposed that four museum specimens from the Seychelles represented a new species which he named *D. arnoldi* (Bour 1982). This discovery prompted Gerlach & Canning (1998a, b) to produce a taxonomic revision of *Dipsochelys*. They now suggested the presence of six species: *D. dussumieri* from Aldabra, *D. hololissa* and *D. arnoldi* from the Seychelles, *D. daudinii* (probably) from the Seychelles, and *D. abrupta* and *D. grandidieri* from Madagascar. Their distinction was also supported in a study on shell morphology by Gerlach (1999) who found notable differences in how the neural bones were arranged (these are the bones that, together with the nuchal, suprapygals and pygal, form the midline of the shell). *D. arnoldi* was reported to be the most divergent, with a special saddle-backed morphology allowing it to be a specialist high-browser.

Excepting *D. arnoldi* Bour, 1982, all of these names are old, so the species recognised by Gerlach & Canning (1998a, b) have been resurrected from synonymy and are not 'new' species in the strictest sense of the term. *D. abrupta* was first coined [as *Geochelone abrupta*] by Grandidier (1868), *D. hololissa* [as *G. hololissa*] by Gunther (1877), *D. daudinii* [as *G. daudinii*] by Dumeril & Bibron (1835), *D. dussumieri* [as *G. dussumieri*] by Gray (1831), and *D. grandidieri* [as *G. grandidieri*] by Vaillant (1885). Until recently all of these Indian Ocean species, excepting the Aldabran *D. dussumieri*, were thought to be extinct. As has been reasonably well reported however, both Bour and Gerlach & Canning reported the discovery of live giant tortoises which appeared to belong to some of the *Dipsochelys* taxa they identified. Gerlach & Canning explained how morphological examination and application of randomly amplified polymorphic DNA analysis (or RAPD) supported the idea that both *D. arnoldi* and *D. hololissa* were still extant in captivity. Alleged *D. arnoldi* and *D. hololissa* specimens have now been identified at zoos and wildlife parks in Mauritius, the Seychelles, Kansas, Michigan, Hawaii and the UK. This is pretty exciting stuff and has prompted a major captive breeding program.

Alarm bells concerning the status of these supposed rediscovered taxa began to ring, however, when Palkovacs *et al.* (2002) could not find any differences among the purported living *Dipsochelys* species in the examined regions of mtDNA, and Palkovacs *et al.* (2003) later produced another study shedding further light on this issue (I should note that Justin Gerlach is on the authorship of both of these papers, so don't go thinking that he's in disagreement with this work). By studying the mtDNA of 55 captive *Dipsochelys* tortoises (including seven specimens identified on morphological criteria as *D. arnoldi* and ten identified as *D. hololissa*), they found no evidence for genetic differentiation between extant *D. dussumieri* individuals and alleged *D. arnoldi* and *D. hololissa* individuals (the initial result of the RAPD analysis reported by Gerlach & Canning was put down to shortcomings with this method and the preliminary nature of the study).

Fig. 25. An Aldabran giant tortoise photographed on La Digue Island. Photo by Evgenia Kononova.

Instead, the results indicate the survival of just a single lineage of Indian Ocean giant tortoises. This contrasts strongly with results obtained from other giant tortoise island radiations (Galapagos tortoises and extinct Mascarene tortoises) where genetic variation between both the populations of different islands, and populations living on the same islands, is high. Given that genetic variation among surviving *Dipsochelys* individuals is so low – and not consistent with the survival of three species – what does this mean? Palkovacs *et al.* (2003) offered three possibilities.

The first is that *D. arnoldi* and *D. hololissa* are extinct and *D. dussumieri* is the only extant *Dipsochelys* after all. This would mean that the morphological criteria on which extant supposed *D. arnoldi* and *D. hololissa* have been identified are utterly unreliable: an idea which matches suggestions that "carapace morphology is sensitive to environmental conditions and that captivity can result in aberrant morphologies" (Palkovacs *et al.* 2003, p. 1409; see also Gerlach 2004b).

The second possibility is that *D. arnoldi* and *D. hololissa* never existed as distinct taxa at all, and the differentiation of these species from *D. dussumieri* was and is unwarranted. As Palkovacs *et al.* (2003) noted, this would only be likely if Aldabran and Seychelles tortoises had not had the opportunity to diverge genetically: therefore, the populations must either have diverged very recently, or there must have been regular and substantial gene flow between Aldabra and the Seychelles. The latter option appears unlikely as it's hard to imagine that many tortoises swimming to and fro between the islands (and, yes, when at sea giant tortoises really do swim). Rapid divergence of phenotype but not of genotype is theoretically possible, but given the genetic closeness of the sampled individuals (and this assumes that the extant Giant tortoises identified as *D. arnoldi* and *D. hololissa* really are remnants of the same populations as those given these names but based on old material), should they be recognised as species separate from *D. dussumieri*? I suppose this would be down to personal taste – that is, whether splitting or lumping is adopted. The third possibility is that hybridisation between Aldabran and Seychelles tortoises has obscured the true genetic distinctiveness of the *D. arnoldi* and *D. hololissa* individuals. In other words, *D. dussumieri* individuals introduced to the Seychelles may conceivably have swamped the mtDNA haplotypes of *D. arnoldi* and *D. hololissa*, even though the unique Seychelles morphology could persist. This idea can be discounted because the suspected hybrids would reveal distinctive microsatellite alleles not occurring in non-hybrid *D. dussumieri*, and they don't.

In conclusion then, Palkovacs *et al.* (2003) showed that there almost certainly aren't extant individuals of *D. arnoldi* and *D. hololissa*. So, assuming that the museum type material of *D. arnoldi* and *D. hololissa* really does represent taxa distinct from *D. dussumieri*, does this mean that all the identifications of living tortoises as representatives of *D. arnoldi* and *D. hololissa* have been incorrect? Or does it mean that *D. arnoldi* and *D. hololissa* were never good species

Fig. 26. A Seychelles giant tortoise supposed to represent *D. arnoldi*.

at all, but that they just represent ecomorphs of *D. dussumieri*?

Genetic data from fossils would help clear up the picture, as would convincing evidence that the living individuals identified as *D. arnoldi* and *D. hololissa* really do represent exactly the same thing as the museum specimens that bear these names. Needless to say these conclusions put doubt on the aims of the captive breeding program, and as yet I haven't heard of a response from any of the people involved (though, as noted above, I haven't seen Gerlach's 2004 book on Indian Ocean tortoises).

Refs - -

- Arnold, E. N. 1979. Indian Ocean giant tortoises: their systematics and island adaptations. *Philosophical Transactions of the Royal Society of London B* 286, 127-145.
- Bour, R. 1982. Contribution a la connaissance des tortues terrestres des Seychelles: definition du genre endemique et description d'une nouvelle probablement originaire des iles granitiques et au bord de l'extinction. *Comptes Rendu de l'Academie des Sciences, Serie III, Sciences de la vie, Paris* 295, 117-122.
- Gerlach, J. 1999. Distinctive neural bones in *Dipsochelys* giant tortoises: structural and taxonomic characters. *Journal of Morphology* 240, 33-37.
- Gerlach, J. 2004a. The complex nasal structure of *Dipsochelys* and its identification as a true Jacobson's organ. *Herpetolological Journal* 15, 15-20.
- Gerlach, J. 2004b. Effects of diet on the systematic utility of the tortoise carapace. *African Journal of Herpetology* 53, 77-85.
- Gerlach, J. & Canning, K. L. 1998a. Taxonomy of Indian Ocean tortoises (*Dipsochelys*). *Chelonian Conservation and Biology* 3, 3-19.
- Gerlach, J. & Canning, K. L. 1998b. Identification of Seychelles giant tortoises. Linnaeus fund research report. *Chelonian Conservation and Biology* 3, 133-135.
- Palkovacs, E. P., Gerlach, J. & Caccone, A. 2002. The evolutionary origin of Indian Ocean tortoises (*Dipsochelys*). *Molecular Phylogenetics and Evolution* 24, 216-227.
- Palkovacs, E. P., Marschner, M., Ciofi, C., Gerlach, J. & Caccone, A. 2003. Are the native giant tortoises from the Seychelles really extinct? A genetic perspective based on mtDNA and microsatellite data. *Molecular Ecology* 12, 1403-1413.

Chapter 8:
Eagle owls take over Britain

Popular mythology has it that the introduction of the 1976 Dangerous Wild Animals Act led to the mass release of numerous pet leopards, pumas and god knows what else into the British countryside, and it's these former pets that today haunt our moors and wooded areas. That's a subject for another time, but right now a very similar subject, but concerning a very different kind of animal, is getting lots of coverage in both the popular and zoological press: the apparent presence in the British countryside of numerous feral Eurasian eagle owls *Bubo bubo*. A formidable bird, it can reach 4 kg with a wingspan of 1.5 m and a length of over 70 cm (Fig. 27). What's this all about, and what's the big deal anyway?

Fig. 27. A Eurasian eagle owl: a formidable bird.

Officially, eagle owls are not British natives. Well, ok, actually they are on the 'British list', but then so are Magnificent frigate birds and Red-billed tropic birds (basically, any bird seen within the boundaries of the British Isles becomes a member of the 'British list'). If they do occur here today, it's as vagrants from continental Europe. So the apparent presence of eagle owls here right now must result from accidental escape, or deliberate release. Unlike the case with British big cats, the existence of feral eagle owls is not doubted by officialdom. They are here: it's official.

The question now is... should we 'keep' them, or should we make efforts to get rid of them? Herein lies the debate. We know without doubt that Britain had eagle owls in the recent past as they're known from palaeontological and archaeological samples. To be sure on this I checked the literature. My usual port of call on British Pleistocene vertebrates is Stuart (1974), but he doesn't list bird taxa. However, specific British *Bubo bubo* specimens from the Pleistocene were described by Harrison (1979, 1987). Intriguingly, there's a new spin on this, mentioned by Jim Giles (2006) in his *Nature* article on the British eagle owl problem. According to Giles, John Stewart (of University College London) has data indicating that eagle owls survived here for longer than had been thought before: "ornithologists had previously assumed that remains [from the past 10,000 years] came from tame eagle owls that had been imported for hunting". Inevitably people have compared this with the recent discovery that lynxes, similarly, were around in Britain until just a thousand or so years ago. If Stewart has data indicating survival of British eagle owls right up to (say) the dawn of the agricultural revolution, he hasn't (to my knowledge) published it. Altogether it strengthens the case for eagle owls being regarded as 'rightful' members of the modern British avifauna, and if we have them back here by accident… well, that's ok.

But should we have them back here? This is the problem. Ok, it's all very nice, but arguments invoking the former presence of a taxon within a region are never particularly convincing because, after all, pretty much all of the Northern Hemisphere's megafauna naturally inhabited Britain at some time within recent geological history. So while I welcome the idea of having eagle owls back in our fauna, I feel that it's problematical, given that the reappearance of any arch predator is dependent on how intact the rest of the food chain is. It's as if we're hoping that we can reconstruct the country's ancient ecosystem by building from the top down. There is lots of talk of reintroducing lynxes and wolves, but not so much talk going on about boosting the numbers and diversity of ground-dwelling slugs, rodents or passerines. Yet it follows that the trophic pyramid that will hypothetically support these arch predators must be reconstructed at its lower levels before things further up are going to fit in nicely. Rabbits. Yes, nowadays we have lots of rabbits that make up a lot of biomass, and they weren't here when eagle owls and lynxes and wolves were truly native, so that might repair part of the problem (rabbits are not native to Britain and were introduced by the Normans). Indeed, some studies find eagle owls to prefer rabbits to all other prey (Hume 1991). If this is right, then things maybe aren't so bad after all. In fact we have too many rabbits as it is, so more predation of them is welcome.

However… other studies find eagle owls to predate mostly on birds (Bocheński *et al.* 1993) with some studies finding 83% of eagle owl diet (by weight) being made up by avian prey (Everett 1977). And even if this weren't the case (and the owl population was still mostly concentrating on rabbits), there is still the fact that eagle owls are opportunistic predators that will still kill animals we don't want them to. So, while – back when eagles owls were healthily distributed natives – they could take their pick of whatever, without this being a problem, things today aren't so rosy, and those potential prey species themselves are depleted in numbers, or even endangered or requiring protection. Indeed the RSPB has voiced concerns that eagle owls may start to impact significantly on Black grouse *Tetrao tetrix*, Hen harriers *Circus cyaneus* and other endangered species.

You see, perhaps the most interesting thing about eagle owls is that they are experts at intraguild predation: in other words, very very good at killing other raptors, and in fact at virtually eliminating them from an area. They routinely take Long-eared owl *Asio otus*, Goshawk *Accipiter gentilis*, Sparrowhawk *A. nisus*, Peregrine *Falco peregrinus*, Gyrfalcon *F. rusticolus*, Merlin *F. columbarius* and Rough-legged buzzard *Buteo lagopus*. More exceptional are cases of predation on Snowy owls *Bubo scandiaca* (note: no longer in its own genus), young White-tailed eagles *Haliaeetus albicilla*, and other eagle owls. Everett (1977) wrote that "up to 5% of [eagle owl] total prey may consist of other birds of prey and ... these may make up as much as 36% of all the bird food consumed in some regions" (p. 93). It seems that eagle owls take these birds while they are roosting, mostly (I assume) by sneaking up on them from behind. In fact, so significant are eagle owls on other raptors that some populations of Long-eared owls appear to migrate specifically because of eagle owl predation (Erritzoe & Fuller 1998).

I don't need to tell you that, right now, Britain does *not* have raptor populations capable of dealing with this sort of predation. Like it or not, the presence of this species is bad news for the raptors we have. So the bottom line is that, while having eagle owls back in Britain is very nice, it is not good news for our native fauna, given the state it's currently in, and in an ideal world, we would need to have the rest of the ecosystem restored before we could bring in the arch predators. However, all of this is academic given that it's already too late. The eagle owls are here, and the native raptors will suffer.

British eagle owls: an update

While at my friend Bernie's house I noticed that the February edition of *Bird Watching* magazine – which I often look at in the shops but never buy. It includes an article by Adrian Thomas on Eagle owls *Bubo bubo* in Britain. The cover states "What is Europe's top hunter doing in the UK?". An article on British eagle owls has also recently appeared in *The Palaeontological Association Newsletter* (McGowan 2006) and several web sites, including those of the RSPB and World Owl Trust (WOT), have also provided new, updated information on this subject within recent months.

The best known British eagle owls are a pair in the Yorkshire Dales that successfully bred on Ministry of Defence land in the spring of 2005. They've actually been breeding since 1997, or possibly 1996, and have managed to raise 23 chicks during this time. Thomas (2006) mentioned "two other confirmed breeding pairs in the UK" (p. 13) and noted that there are possibly more, and it has also been brought to attention that pairs in Galloway, Invernesshire and Sutherland have also been confirmed as breeders.

Fig. 28. Another Eurasian eagle owl. Despite its size and habits, it is not a bird of wild, remote regions, but may often live close to people.

Already some of the chicks of the Yorkshire pair have moved far afield, with one of them having been reported from Shropshire (where it was electrocuted on power lines). As Tommy Tyrberg has noted, eagle owls aren't really birds of "surviving vestiges of wilderness, immune from human exploitation" as it says in *Birds of the Western Palearctic*, but are actually quite happy living close to people. Populations in The Netherlands and Sweden are doing ok close to noisy quarries, in working farmland, and in and around rubbish dumps. Britain is a small place with no wilderness at all, and evidence for humans and their recent activity is everywhere, so the adaptability of the eagle owl, and its success on the continent despite human activity, certainly suggests that it'll do fine in this country.

Having said that, in January 2006 came the disappointing news that the female of the Yorkshire pair had been killed, with its death apparently occurring just before Christmas. Autopsy showed that the shot used was large-gauge like that used to shoot foxes, and not small-gauge like that used for hunting gamebirds (thus probably ruling out a case of mistaken identity). The bird had an empty stomach, and thus may have starved to death after being shot. If eagle owls are colonising our islands naturally, this is a sad loss of an important individual. The killing may also have been illegal – ultimately this depends on whether or not the bird was here 'naturally' – and the North Yorkshire Police are pursuing enquiries.

Thomas (2006) also provided some new discussion on the source of origin of the British eagle owls: might they be vagrants that are naturally colonising Britain from continental Europe? He noted two pieces of evidence that might support this hypothesis. Firstly, the television documentary *Return of the Eagle Owl* looked at the research of raptor conservationist Roy Dennis. His examination of 18th and 19th century eagle owl records in Britain shows that the birds were mostly reported between September and January – the time "when one might expect vagrants to arrive" (Thomas 2006, p. 15).

Fig. 29. A female Eurasian eagle owl. Photo by Adam Kumiszcza.

Secondly, ornithologists monitoring eagle owls in Switzerland have shown that the birds can move as far as 350 km, passing obstacles such as major mountain ranges as they go. A hop across the English Channel may therefore seem no trouble at all, and indeed we know that some European owls, like Long-eared owls *Asio otus* and Short-eared owls *A. flammeus*, cross bodies of water like the North Sea regularly.

Thomas (2006) countered that sedentary European birds seem to find the English Channel and/or the North Sea an insurmountable barrier: Black woodpeckers *Dryocopus martius* don't cross the channel for example, even though their range approaches the French coastline. Similarly,

Eurasian pygmy owls *Glaucidium passerinum* – widespread in Scandivania where they range right up to the North Sea coastline – have never colonised Britain. Ural owls *Strix uralensis* and Great grey owls *S. nebulosa* are widespread in Sweden, but also haven't colonised Britain. Tawny owls *S. aluco* don't cross the Irish Sea.

However, just because these species are sedentary doesn't mean that eagle owls have to be too. As the Long-eared and Short-eared owls show, rules on dispersal capability vary among species and there isn't a single rule that applies to all Strigidae. The possibility that British eagle owls are natural colonisers is therefore worthy of consideration and needs more investigation.

Tony Warburton of the World Owl Trust has written a piece on the WOT's position. They are confident that the British eagle owls are natural colonisers, and that confirmation of eagle owl breeding in Britain is news akin to that of the reintroduction of the White-tailed eagle *Haliaeetus albicilla* or the successful increase in Red kite *Milvus milvus* numbers. The WOT also contends that the British Ornithologists' Union should now add Eurasian eagle owl to the official British bird list and that they should receive full protection.

What does the BOU say about this? In a 1996 review of the eagle owl's status in Britain, they concluded that insufficient evidence was available to accept the species on the British list (they concluded that the 90 reports they examined were either not definitely of *Bubo bubo*, or might have been of birds that had escaped from captivity). So far as I can tell, they are watching the situation but are not yet prepared to be as positive as the WOT is about possible native status. Similarly, the RSPB is being cautious: they say that they would be more than happy to accept the species as a native, but compelling evidence that demonstrates this has yet to be produced.

We have some more news on fossils. I noted earlier that Giles (2006) drew attention to John Stewart's mention of possible post-glacial eagle owl fossils. In *The Palaeontological Association Newsletter* article mentioned above, Al McGowan (2006) also mentioned Stewart's interest in this subject and, even better, discussed and figured an eagle owl carpometacarpus from post-glacial deposits near Cheddar, Somerset. This provides powerful support for the natural presence of the species in modern Britain ('modern' in the geological sense you understand).

Of course, if our eagle owls have gotten here naturally, they would be protected under the Wildlife and Countryside Act. Any impact that they have on other British animals – and as discussed above they *might* affect raptor numbers as well as those of Black grouse – will be something we can record, but not intervene in. Conversely, if it can be shown that the British eagle owls were bred in captivity and later released, they don't deserve protection and an argument could be made that they should be removed.

Given the balance of evidence, my feeling at the moment is that at least some of our eagles owls *have* gotten here themselves.

We await further news.

Refs - -

- Bocheński, Z., Tomek, T., Boev, Z. & Mitev, I. 1993. Patterns of bird bone fragmentary in pellets of the Tawny owl (*Strix aluco*) and the Eagle owl (*Bubo bubo*) and their taphonomic implications. *Acta zool. cracov.* 36, 313-328.
- Erritzoe, J. & Fuller, R. 1998. Sex differences in winter distribution of Long-eared owls (*Asio otus*) in Denmark and neighbouring countries. *Vogelwarte* 40, 80-87.
- Everett, M. 1977. *A Natural History of Owls*. Hamlyn, London.
- Giles, J. 2006. Bird lovers keep sharp eye on owls. *Nature* 439, 127.
- Harrison, C. J. O. 1979. Birds of the Cromer Forest Bed series of the East Anglian Pleistocene. *Transactions of the Norfolk and Norwich Naturalists' Society* 24, 277-286
- Harrison, C. J. O. 1987. Pleistocene and prehistoric birds of south-west Britain. *Proceedings of the University of Bristol Spelaeology Society* 18, 81-104.
- Hume, R. 1997. *Owls of the World.* Parkgate Books, London.
- McGowan, A. 2006. Should eagle owls be considered native to the UK? *The Palaeontological Association Newsletter* 61, 21-23.
- Stuart, A. J. 1974. Pleistocene history of the British vertebrate fauna. *Biological Reviews* 49, 225-266
- Thomas, A. 2006. Where eagles dare. *Bird Watching* Feb' 2006, 12-18.

Chapter 9:

When whales walked the land... and looked like antelopes... and mimicked crocodiles... and evolved trunks. What?

Everybody interested in animals is, I assume, fascinated by whales, and one of the most interesting topics in mammal evolution surely has to be the evolution of whales from their terrestrial ancestors. As a relatively well documented macroevolutionary tale (Thewissen & Bajpai 2001), the transition integral to this story is informative on so many levels (see Carl Zimmer's *At the Water's Edge* for a good, though now dated, overview). For me, the early evolution of whales has become particularly interesting for two reasons. Firstly, it's controversial, and particularly so within recent years. A cherished idea of cetacean affinities – namely, the notion that whales descend from mesonychians – has been usurped by the originally unpopular and crazy notion that whales might be close relatives of hippos.

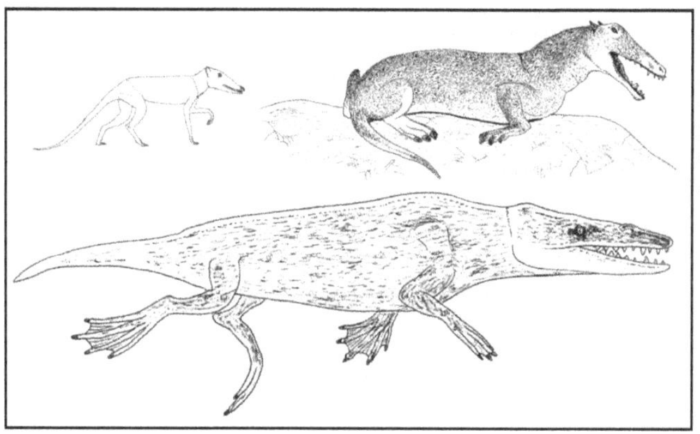

Fig. 30. An assemblage of Eocene stem-cetaceans: a pakicetid is shown at top left, *Ambulocetus* at top right, and the protocetid *Rodhocetus* at bottom.

Secondly, it's an area illuminated by novelty: specifically, by lots of newly recognised taxa. It turns out that Eocene taxa, the so-called archaeocetes, weren't just seal-shaped amphibious protowhales: they were a motley assortment that got up to all sorts of things. Believe it or not, they included saltatorial deer-like forms, macropredatory crocodile-mimics, diminutive long-tailed forms that looked like giant desmans, 'megaseals' and forms sporting tapir-like trunks. Several of the taxa I have in mind are new (as in, published in 2004 or 2005), and already a review article I wrote on basal cetaceans (Naish 2004) is out of date.

Though basal whales less specialised for aquatic life than modern forms had been known since the 1830s (these being basilosaurids and dorudontids), it was only in 1904 that fossil whales truly reminiscent of their terrestrial ancestors were discovered. Two new species from Lutetian (middle Eocene) rocks of Egypt, *Protocetus atavus* and *Eocetus schweinfurthi*, were described by Eberhard Fraas in that year. With their geological antiquity (they were, at the time, the oldest of all whales), relatively small body size (3-4 m, compared to 5-20 m for basilosaurids and dorudontids), and teeth and vertebrae more similar to those of typical land mammals than to those of other whales, they were clearly the most primitive of all whales then known and were deemed worthy of their own new family, Protocetidae Fraas, 1904. However, exhibiting an elongate rostrum and uniquely shaped dense-boned ear capsules, they were clearly still members of Cetacea. Numerous discoveries of fossil whales similar to *Protocetus* and *Eocetus* have since shown that protocetids and their relatives experimented with a variety of lifestyles in and adjacent to the marine environment. Discoveries of *Protocetus*-like material in North America (Kellogg 1936) also show that these animals were geographically widespread.

During the 1970s and 80s emphasis in basal whale research began to shift from Africa to Asia, at first because Sahni & Mishra (1972) described primitive whale remains discovered in India, and later as West (1980) figured and identified lower jaw specimens from Pakistan. Sahni & Mishra (1972) regarded their material as representing two new species of *Protocetus*, *P. sloani* and *P. harudiensis*, the new taxon *Indocetus ramani*, and two new basal odontocetes, *Andrewsiphius kutchensis* and *A. minor*. Their *Protocetus* species have since been shown to be referable to a new genus that is part of a newly recognised, highly unusual basal whale group, Remingtonocetidae (Kumar & Sahni 1986). *Andrewsiphius* also proved to be a remingtonocetid. West (1980) also named a Pakistani specimen as a new species of *Protocetus* – *P. attocki* (but see below) – and also formalised previous suggestions that two tooth-based Pakistani mammals, *Ichthyolestes pinfoldi* and *Gandakasia potens*, were not mesonychians, as they had been originally described, but basal whales. By now, the number of Asian basal cetacean taxa had exceeded the African count.

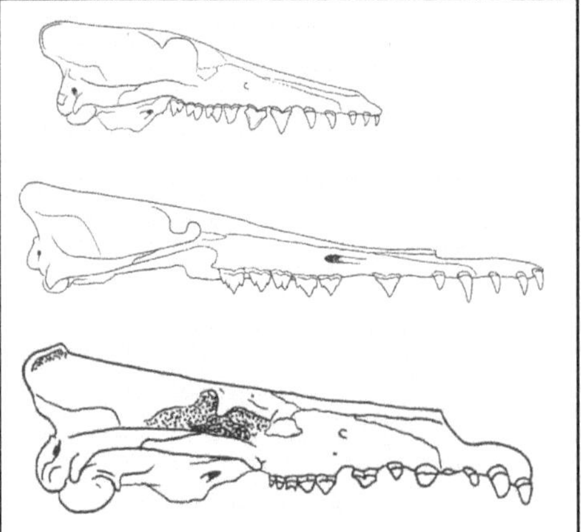

Fig. 31. Diverse skulls of Eocene whales. From top to bottom: *Pakicetus*, *Remingtonocetus* and protocetid *Artiocetus*. Not shown to scale.

The next most significant basal cetacean discovery was announced in 1981 when Gingerich & Russell (1981) described the well preserved but incomplete skull of the new genus and species *Pakicetus inachus*. West's *P. attocki* proved to be a second species. With a number of unique cetacean ear bone characteristics, *Pakicetus* was interpreted as a probably amphibious predator that was transitional between terrestrial, wolf-like mesonychians, the probable ancestors of whales (or so it was thought at the time), and larger, younger archaeocetes like the basilosaurids. The ear bones of *Pakicetus* are uniquely intermediate between those of land mammals and those of derived cetaceans (Gingerich *et al.* 1983) and, coupled with its presence in shallow freshwater environments, suggest that it was not yet a fully marine animal.

With only a partial skull to go on, *Pakicetus* was assumed to be seal-like, and conventional life restorations depict it that way. We now know that these restorations are wildly inaccurate, as associated pakicetid skeletons show that these mammals weren't all that different, superficially, from basal 'slinker-type' artiodactyls such as chevrotains, small deer and basal Eocene forms (Thewissen *et al.* 2001). In fact we also now know that pakicetids had the double-pulley astragalus regarded as unique to artiodactyls, so now it seems that artiodactyls and cetaceans were closer to one another than cetaceans were to mesonychians. Far from being seal-like protowhales, pakicetids – the most basal members of the cetacean radiation – were more like carnivorous little antelopes, though presumably they spent at least some time in shallow water.

After *Pakicetus*, perhaps the most important basal whale discovery was that of *Ambulocetus natans*, a Pakistani whale a few million years younger than *Pakicetus* (Thewissen *et al.* 1994). With a dorsoventrally mobile back and huge, long-toed feet, *Ambulocetus* appears suited for competent movement on land, but probably swam by powerful up-and-down oscillation of the back and paddling with the hindlimbs. Its elongate tail was not powerfully muscled and does not have the special square-shaped vertebrae required to support a tail fluke.

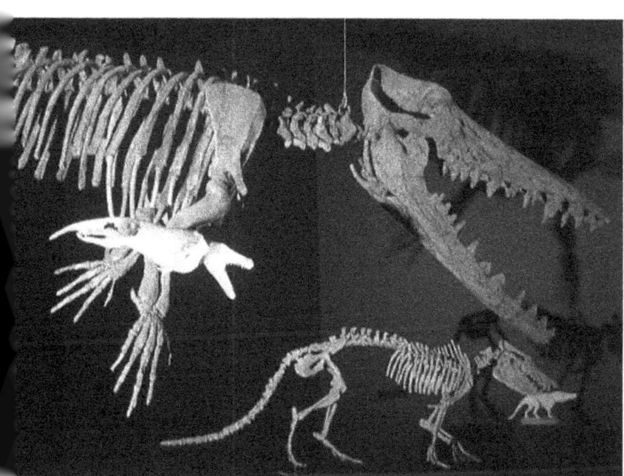

Fig. 32. A reconstructed skeleton of *Ambulocetus*, shown 'swimming' over a reconstructed pakicetid skeleton.

In contrast to pakicetids, *Ambulocetus* had a narrow snout and laterally directed orbits, relatively short legs, and an overall robust skeleton. An intriguing possibility regarding the palaeobiology of *Ambulocetus* is that it was a crocodile-like ambush predator that stalked prey while concealed in shallow water (Thewissen *et al.* 1996). This conclusion is based on the crocodile-like skull of *Ambulocetus*: both groups have long but robust snouts, pointed teeth, strong jaw-closing muscles and eyes located high up on the head.

Ambulocetus also exhibits unusual and extensive tooth wear which indicates that it fed on bony prey. If the crocodile analogy is viable, what might *Ambulocetus* have preyed on? Anthracobunids, a group of herbivorous ungulates probably related to elephants, and sirenians both lived in the same environments as *Ambulocetus*. Thewissen *et al.* (1996) suggested that *Ambulocetus* might have preyed upon these animals, as well as other animals that may have approached the shoreline.

However, given that large crocodiles were already common in the marine and freshwater environments frequented by *Ambulocetus*, this theory has a flaw. Indeed, it has been pointed out before that primitive cetaceans may have suffered from extensive crocodilian competition and predation. Perhaps, therefore, *Ambulocetus* selected environments where crocodiles were very rare or absent.

The discovery of *Ambulocetus* and other new basal whales has shown that they can't all be shoehorned into an all-encompassing Protocetidae. Rather, they should be classified as several distinct groups. *Pakicetus* and its close relatives were given their own group, Pakicetinae, by Gingerich & Russell (1981), and this has more recently been raised to the level of 'family'. Besides *Pakicetus*, *Ichthyolestes pinfoldi* and *Nalacetus ratimitus* are included. Thewissen *et al.* (1996) coined Ambulocetidae for *Ambulocetus*, and it seems that *Gandakasia* and *Himalayacetus* are also members of this group.

Remingtonocetidae was named by Kumar & Sahni (1996) for *Remingtonocetus* and *Andrewsiphius*, two very peculiar long-snouted middle Eocene whales from India and Pakistan. *Andrewsiphius* had previously been regarded as an odontocete on the basis of its elongate and compressed lower jaw (Sahni & Mishra 1975). However, this referral was not defensible as, like archaeocetes and unlike odontocetes, *Andrewsiphius* had a primitive style of tooth count, relatively deep lower jaws, and intermittent gaps between the teeth. Remingtonocetids are also odd in having a particularly long, slender snout and widely set small eyes. The most informative of them is little *Kutchicetus minimus* (Bajpai & Thewissen 2000), the smallest known Eocene whale (though note that not all early whale experts agree that it is distinct from *Remingtonocetus)*. It shows that remingtonocetids had short, stout limbs, a long, sinuous body and a remarkably long tail. I like to think that it looked like a giant desman, but that only helps if you know what a desman is.

Refs - -

- Bajpai, S., Thewissen, J. G. M. 2000. A new, diminutive Eocene whale from Kachchh (Gujarat, India) and its implications for locomotor evolution of cetaceans. *Current Science* 79, 1478-1482.
- Gingerich, P. D. & Russell, D. E. 1981. *Pakicetus inachus*, a new archaeocete (Mammalia, Cetacea) from the Early-Middle Eocen Kuldana Formation of Kohat (Pakistan). *Contributions from the Museum of Paleontology, University of Michigan* 25, 235-246.
- Gingerich, P. D., Wells, N. A., Russell, D. E. & Ibrahim Shah, S. M. 1983. Origin of whales in epicontinental remnant seas: new evidence from the Early Eocene of

Pakistan. *Science* 220, 403-406.

- Kellogg, R. 1936. A review of the Archaeoceti. *Carnegie Institute of Washington Publication* 482, 1-366.
- Kumar, K. & Sahni, A. 1986. *Remingtonocetus harudiensis*, new combination, a Middle Eocene archaeocete (Mammalia, Cetacea) from western Kutch, India. *Journal of Vertebrate Paleontology* 6, 326-349.
- Naish, D. 2004. Fossils explained 46. Ancient toothed whales. *Geology Today* 20 (2), 72-77.
- Sahni, A. & Mishra, V. P. 1972. A new species of *Protocetus* (Cetacea) from the Middle Eocene of Kutch, western India. *Palaeontology* 15, 490-495.
- Thewissen, J. G. M. 1998. Cetacean origins. Evolutionary turmoil during the invasion of the oceans. In Thewissen, J. G. M. (ed) *The Emergence of Whales*. Plenum Press (New York), pp. 451-464.
- Thewissen, J. G. M. & Bajpai, S. 2001. Whale origins as a poster child for macroevolution. *BioScience* 51, 1017-1029.
- Thewissen, J. G. M., Hussain, S. T. & Arif, M. 1994. Fossil evidence for the origin of aquatic locomotion in archaeocete whales. *Science* 263, 210-212.
- Thewissen, J. G. M., Madar, S. I. & Hussain, S. T. 1996. *Ambulocetus natans*, an Eocene cetacean (Mammalia) from Pakistan. *Courier Forschungsinstitut Senckenberg* 191, 1-86.
- Thewissen, J. G. M., Williams, E. M., Roe, L. J. & Hussain, S. T. 2001. Skeletons of terrestrial cetaceans and the relationship of whales to artiodactyls. *Nature* 413, 277-281.
- West, R. M. 1980. Middle Eocene large mammal assemblage with Tethyan affinities, Ganda Kas region, Pakistan. *Palaeontology* 54, 508-533.
- Zimmer, C. 1998. *At the Water's Edge: Macroevolution and the Transformation of Life*. Free Press (New York), pp. 290.

Chapter 10:
British big cats: how good, or bad, is the evidence?

As someone trying to gain a reputation as a credible scientist, it is not in my interest to declare my fascination with alien big cats (as they're known: ABCs from hereon). This is generally regarded, especially in academic circles, as a crackpot area inhabited only by the lunatic fringe. Unfortunately, this stigma – accentuated by the half-serious, sensationalised way the subject is treated by journalists – has tarnished what is actually a perfectly sensible area for which good scientific data exists. When analysed by qualified scientists (whether they be field ecologists, laboratory-based specialists, or image analysts, or whatever), the results have been mostly positive. I start my talks on this subject by emphasizing that I do not 'believe' in ABCs (meaning that I do not accept their reality without question, as this is what is meant by the term 'believe' – see Arment (2004) for more on that if you're interested). Furthermore, I have tried my best to maintain an appropriately sceptical approach. Like any scientist approaching a problem, I have come to the conclusions that I have because that is where the evidence has led me.

And having become acquainted with the large amount of data, I have a dilemma. On the one hand I feel that the data is so compelling that we should accept ABC reality without question, and proceed with the realisation that ABCs are undoubtedly genuine. But on the other hand I feel that more, and better, evidence is required for us to be so confident. Let me make this clear: the evidence is outstanding, and none of the doubts expressed about this subject have any standing. It's often said that, if ABCs are real, then why don't we have good photos, why don't we have dead bodies, why don't we have captured live animals, and why don't we have definitive track and sign evidence? Well, the news is that we do have good photos, we do have dead bodies, we do have captured live animals, and we do have definitive track and sign evidence. This data is out there for anyone that's prepared to examine it. Why isn't this more widely known? That's the mystery. The negative stigma attached to the subject seems to mean that the good data doesn't really get out, at least to those people who haven't gone to the trouble of immersing themselves in the subject.

Fig. 33. Captive Scottish wildcat.
Photo by Neil Phillips.

Before I continue I should add that Britain only has, officially, two native felids: Scottish wildcats *Felis silvestris* (Fig. 33) and Kellas cats. There is considerable disagreement as to whether the former should be kept as a distinct species: if Scottish wildcats are conspecific with the domestic cat *F. catus*, then African wild cats *F. lybica* and Indian desert cats *F. ornata* should be too – they're even closer to *F. catus*, and indeed *F. lybica* is probably ancestral to domestics. Indeed it is even doubted by some as to whether purebred wildcats exist in Britain anymore. Others, however, regard it as useful to keep *F. silvestris* as distinct (see French *et al.* 1988, Daniels *et al.* 1998, Kitchener 1998, Reig *et al.* 2001, Pierpaoli *et al.* 2003). Kellas cats, only discovered in 1984, are introgressive domestic cat x wildcat hybrids that appear to be evolving their own unique behaviour and morphology and, by inference, into a new species (Shuker 1990).

On to the ABC evidence itself, firstly, there are hairs, tracks and droppings that have been conclusively identified by experts as having come from non-native felids. Cat hairs recovered from a site in Lincolnshire in 2003 were confirmed by a government-accredited laboratory as having come from a member of the genus *Panthera*. Droppings collected in 1993 from Whorlton, County Durham, were identified by Hans Kruuk as from a Puma *Puma concolor*. This is a big deal because Kruuk is a world authority on the field biology and ecology of carnivorans, and he's otherwise been openly sceptical of the existence of ABCs. Finally, a large number of trackways from various locations across the UK seem to be big cat tracks. That is, they possess the diagnostic features seen in cat tracks, but not in those of dogs and other carnivorans. It is, however, admittedly difficult to be really sure on tracks, and many of the alleged ABC tracks that I've seen – while probably produced by cats – leave room for doubt.

It's well known that huge number of livestock kills have been blamed on ABCs. It's often said that the way an animal has been killed – the apparent 'neatness' of the wound and resulting feeding sign, and the fact that the animal seems to have been killed by an attack to the throat – is indicative of a big cat as killer. This might be valid, but it's very difficult to be confident about, so I hesitate in regarding livestock kills as that informative. There is, however, one specific case that stands out head and shoulders above the others: the Cupar roe deer carcass. There is no doubt in my mind that this animal was killed by a big cat, and it was found (by Ralph Barnett, a journalist with no prior interest in the ABC phenomenon) on a small country road in Scotland. Full discussion of this case is given below.

Similarly compelling are dead bodies. Yes, dead bodies of British ABCs. Several specimens are now known from the UK, and they show that several species of exotic felids are (at least at times) abroad in the British countryside. They include several Jungle cats *Felis chaus*, five Leopard cats *Prionailurus bengalensis*, and a Eurasian lynx *Lynx lynx*, shot dead in East Sussex in 1991 (Shuker 1995). The lynx is particularly interesting as the case was pretty much kept quiet until 2001. I don't hold much faith in conspiracy theories, but the farmer who shot this animal was told by the police to keep it to himself. In fact, it's not difficult to think that, if any official body (say, the police, or the government) does know that ABCs are an undoubted reality, they will likely not want this to become widely known. Two live exotic felids have been captured in Britain: a puma called Felicity, and a lynx called Lara. Both animals have been regarded as escapees from private collections, but this is missing the point given that surely all British ABCs are escapees from private collections (read on).

Fig. 34. Still image from a sequence of film showing a large dark cat filmed in Worcestershire in 1992.

Finally, an increasing number of still photos and sequences of video footage are definitive and clearly show British ABCs. They also depict assorted species. An excellent photo taken by Peter Nixon at County Durham in 1992 clearly shows a Jungle cat, a black leopard was photographed a few times as it ran across a hillside at Tonmawr, Wales, by Di Francis in 1982, and a large cat that is either a puma or black leopard was photographed in November 1988 by Tim Young at Zennor, Cornwall (this photo is on the cover of Nigel Brierley's *They Stalk By Night*). My favourite images come from a short sequence of video footage filmed at Great Witley, Worcestershire, by Nick Morris in May 1992 (Fig. 34). The animal is a black leopard (though this is only really clear when the original colour footage is viewed). There are other bits of excellent, definitive photographic evidence – those I've mentioned are my favourite ones.

So what does all this mean? The conclusion is that the evidence for the presence of exotic felids in the British countryside is overwhelmingly good, and it can't seriously be doubted that the animals are here. The Cupar roe deer carcass, definitive hairs, droppings, outstanding photographic data, and dead bodies and captured animals, provide compelling data for the contention that ABCs are real. Notable efforts to find evidence that resulted in negative conclusions (Baker & Wilson 1995, Weidensaul 2002) did so because, I think, they didn't look at enough data. This was particularly obvious in the case of Baker & Wilson's study (produced for the Ministry of Agriculture Fisheries and Food): they only looked at four pieces of photographic data, for example, and had a total budget for their study of £8200 (Moiser 2001). Exotic cats of several species are here, and in fact the term ABC is a misnomer, as a significant percentage of the animals are not big cats in the strict sense of the word, but members of various small cat lineages. I assume that, because these cats are rather larger than domestic cats, they are therefore assumed to be 'big cats'. That goes even for Jungle cats, which are about a third bigger than a domestic moggie, but no where near the size of a puma or leopard.

The great mystery I suppose is why these cats are here. They are not natives, and even though we now know that lynxes were here until about 1000 years ago, it is implausible that mammals this large could remain undetected in such small, crowded islands. They are escapees from collections, or animals that have been deliberately introduced. Can an exotic felid survive in the British countryside? Yes, without doubt. In fact a Clouded leopard *Neofelis nebulosus* that escaped from Howletts Zoo in 1975 survived for nine months in the wild until it was shot, and it was healthy and in good condition. Yet this is one of the most specialised, tropical cats of them all.

The Cupar roe deer carcass

For many years farmers and other people have reported finding the carcasses of large mammals – mostly sheep but also calves, foals and other livestock – that seem to have been killed by ABCs (for photos see Brierly 1989, Francis 1983, 1993). Supposedly, the wounds present on these corpses, and the manner in which they have been gutted and/or eaten, are diagnostic of felid killers. But like many who have tried to examine this body of evidence impartially, I remain sceptical, and in virtually all cases it is never really clear that dogs can be excluded outright. But there is one exception that stands head and shoulders above all the others: the

Cupar roe deer carcass.

On the night of June 16th 2001, journalist Ralph Barnett was driving home from Dundee to Cupar (north-east Fife, Scotland). As a journalist, Barnett has admitted familiarity with the subject of ABCS, and in particular with the ABCs of Scotland, but he had no special prior interest in the subject. On rounding a bend and coming out of a slight dip in the road, he switched his headlamps to full beam. What he took to be the headlamps of another car immediately ahead caused him to undertake an emergency stop, but it wasn't a car in front of him, it was – so he reports – a big dark-coloured cat. It leapt away out of sight, and as it did Barnett realised that it had been feeding on the carcass of a Roe deer *Capreolus capreolus*, still lying there in the road.

Barnett called the local police on his mobile phone and they "attended in significant numbers – certainly more than would normally be available for a disturbance in Cupar town centre at that time on a Saturday night". The police elected not to retain the carcass and it was unfortunately dumped at the roadside and left there, but Barnett took excellent photos (Figs 35-36). A detailed description of the carcass was posted online to accompany the images, and after being asked questions about the carcass by several ABC investigators Barnett supplied further additional details.

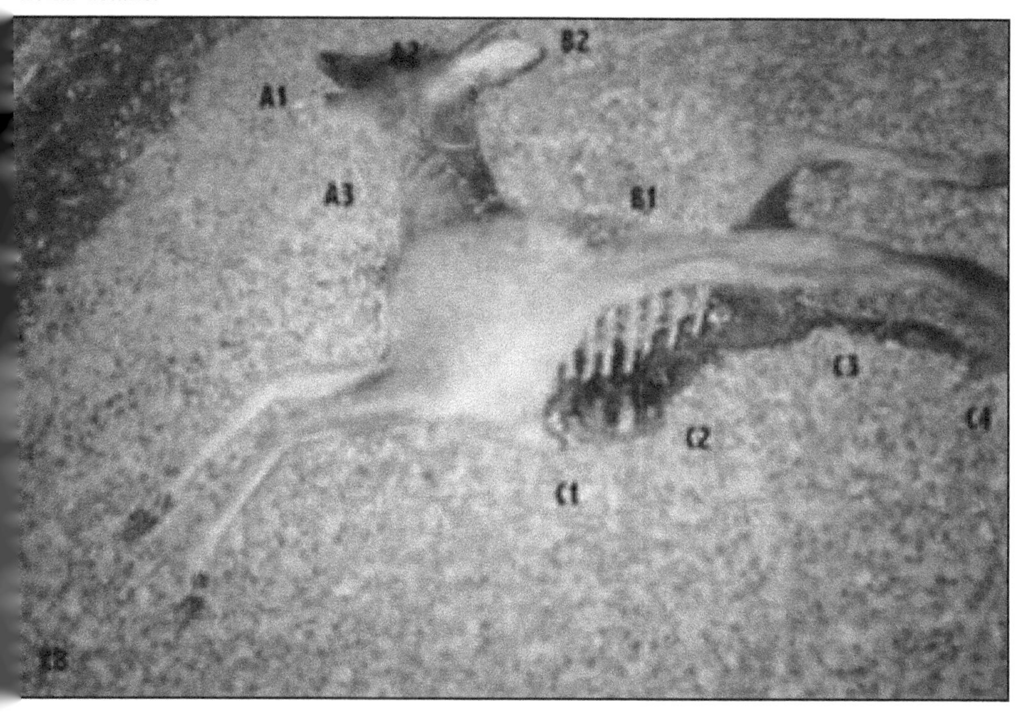

Fig. 35. The Cupar deer carcass. Photo by Ralph Barnett.
Image courtesy of Big Cats in Britain

As seen in the accompanying close-up (Fig. 36), the deer seems to have been killed by asphyxiation. This is evidenced by bulging eyes, an open mouth with protruding tongue and clotted blood pooled on the side of the face. The eyeballs were ruptured and still moist. A series of sub-parallel lacerations on the side of the neck look exactly like claw marks (and were interpreted as such by Barnett): they were deep grooves incised into the neck.

The fact that the carcass was in the middle of the road suggests that it was dragged there (Barnett suggested that the cat was in the process of moving the carcass when he chanced upon it). In keeping with this the carcass had been eviscerated, and what appeared to be a sub-circular grip mark was present on one of its shoulders. The carcass was cold to the touch and without signs of decomposition, and both Barnett and a police officer agreed that it had been dead for less than 48 hours. The tip of one of the antlers was broken off, which would also be in keeping with the carcass having been dragged across the road surface. The entire carcass was split open along its ventral surface, the bones of its pelvis were partially dislocated, and its left hindlimb was defleshed right down to the bones. Its ribs had apparently been cleanly broken. Barnett reported that moist blood, tufts of deer hair and disturbed earth were present at the side of the road.

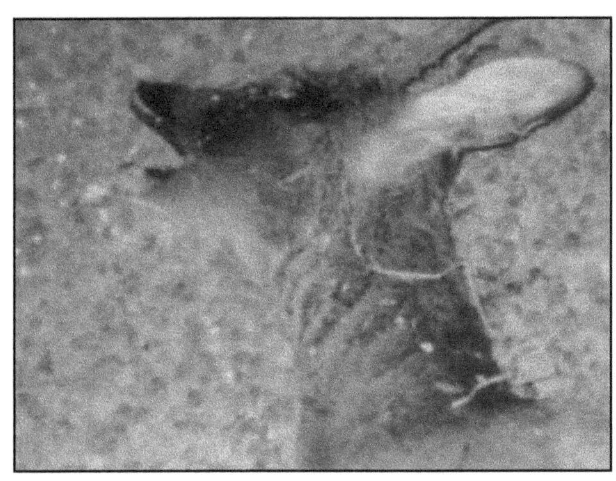

Fig. 36. The head of the Cupar deer carcass. Note the dark area of clotted blood on the face, protruding tongue and marks on the neck. Photo by Ralph Barnett. Image courtesy of Big Cats in Britain

So far as I can tell – and this opinion is echoed by those who have investigated the details provided by Barnett – this is a pretty convincing big cat kill. The extensive trauma present on the carcass simply cannot have been caused by anything else. The good evidence for asphyxiation strongly suggests that the deer was killed by a conventional felid throat-hold: if anyone can come up with a better explanation for bulging, ruptured eyes, a protruding tongue and clotted blood massed on the side of the face I'd like to hear it. The only way you could fake this is by catching the deer live and strangling it to death by hand, and this doesn't strike me as likely.

For me, this case is a big deal as it's the only truly compelling British big cat kill: there are others, sure, but the evidence hasn't been as well documented or reported, nor is it available. Whether Ralph Barnett really encountered a big cat crouching over that carcass is of course something that only he knows, though personally I see little reason to doubt the veracity of his account. However, whether he saw what he said he did or not is irrelevant as the photos speak for themselves. Given that the other lines of evidence we have for British ABCs – the hair, photographic evidence, and the dead bodies – already demonstrate that the animals are a reality, it is inevitable that genuine big cat kills would be discovered and documented eventually. In my opinion the Cupar roe deer carcass is the first good, well documented example, and as such it's an important piece of additional evidence for ABC reality.

Finally, it's worth noting that Roe deer are ideal prey for big cats like pumas and leopards, and they are regularly predated upon by leopards where the two coexist.

Incidentally, Britain's roe deer are usually thought of as native – in fact together with Red deer *Cervus elephus* they are always said to be our only truly native deer (and this is in a country with seven wild deer species). But it's little known that Roe deer were in fact extinct across most of southern Britain by the 18th century and have since been restocked from elsewhere (mostly from Scotland). Nowak (1999) – that's *Walker's Mammals of the World* (Sixth Edition) – cited Christopher Lever's *The Naturalized Animals of the British Isles* (Lever 1977) as the source for this, but Lever only mentions roe deer once and not in connection with this successful reintroduction. An excellent source on the history of roe deer in Britain is Richard Prior's *Living With Deer* (Prior 1965). While a handful of English roe deer *might* be true natives, it's only really those of Scotland that represent the original populations.

Refs - -

- Arment, C. 2004. *Cryptozoology: Science & Speculation.* Coachwhip Publications (Landisville, Pennsylvania), pp. 393.
- Baker, S. J. & Wilson, C. J. 1995. *The evidence for the presence of large exotic cats in the Bodmin area and their possible impact on livestock.* ADAS, Ministry of Agriculture Fisheries and Food, pp. 16.
- Brierly, N. 1988. *They Stalk by Night – the Big Cats of Exmoor and the South West.* Yeo Valley Publications, Bishops Nympton.
- Daniels, M. J., Balharry, D., Hirst, D., Kitchener, A. C. & Aspinall, R. J. 1998. Morphological and pelage characteristics of wild living cats in Scotland: implications for defining the 'wildcat'. *Journal of Zoology* 244, 231-247.
- Francis, D. 1983. *Cat Country.* David and Charles, Newton Abbot.
- Francis, D. 1993. *The Beast of Exmoor and Other Mystery Predators of Britain.* Jonathan Cape, London.
- French, D. D., Corbett, L. K. & Easterbee, N. 1988. Morphological discriminants of Scottish wildcats (*Felis silvestris*), domestic cats (*F. catus*) and their hybrids. *Journal of Zoology* 214, 235-259.
- Kitchener, A. C. 1998. The Scottish wildcat – a cat with an identity crisis? *British Wildlife* 9, 232-242.
- Lever, C. 1977. *The Naturalized Animals of the British Isles*. Hutchinson & Co, London.
- Moiser, C. 2001. *Mystery Cats of Devon and Cornwall.* Bossiney Books, Launceston.
- Nowak, R. M. 1999. *Walker's Mammals of the World, Sixth Edition* (two volumes). The Johns Hopkins University Press, Baltimore and London.

- Pierpaoli, M., Birò, Z. S., Herrman, M., Hupe, K., Fernandes, M., Ragni, B., Szemethy, L. & Randi, E. 2003. Genetic distinction of wildcat (*Felis silvestris*) populations in Europe, and hybridisation with domestic cats in Hungary. *Molecular Ecology* 12, 2585-2598.
- Prior, R. 1965. *Living With Deer.* Andre Deutsch, London
- Reig, S., Daniels, M. J. & Macdonald, D. W. 2001. Craniometric differentiation within wild-living cats in Scotland using 3D morphometrics. *Journal of Zoology* 253, 121-132.
- Shuker, K. P. N. 1990. The Kellas cat: reviewing an enigma. *Cryptozoology* 9, 26-40.
- Shuker, K. P. N. 1995. British mystery cats – the bodies of evidence. *Fortean Studies* 2, 143-152.
- Weidensaul, S. 2002. *The Ghost With Trembling Wings*. North Point Press (New York).

Chapter 11:
Luis Rey and the new oviraptorosaur panoply

Earlier this week (note: the text you're reading was written in February 2006) I spent a few days in London: Mike Taylor and I had arranged to work in the collections of the Natural History Museum on Wealden sauropods. We have what appears to be a new taxon. I also took the opportunity to complete some of my research on large Wealden theropods. On the way, I stopped off and stayed with Luis Rey and his partner Carmen, and it's the fantastic new stuff Luis showed me that I want to discuss here.

Right now Luis is very busy doing the artwork for a major new book on dinosaurs written by Tom Holtz. Judging from the art I've seen, it will be spectacular and one of the most attractive dinosaur books ever. Several plates depict a taxonomic panoply of a particular group, and we also have the first accurate life restoration of the bizarre *Lurdusaurus*, as well as new restorations of *Dilong*, *Guanlong*, *Scutellosaurus* and so many others. What caught my imagination in particular were his restorations of the amazing diversity of recently named oviraptorosaurs (a black-and-white version depicting this diversity has been published before (Gee & Rey 2003), but it did not include as many taxa). In fact there are now so many members of this group that it is proving difficult to keep up, and difficult to keep track of what is what, especially given that several specimens have been incorrectly allocated to a genus and later reallocated or re-named. In fact several discoveries relevant to this area have appeared in recent months and I've only just done reading them, so now is a good time to review it.

Most of the new taxa belong to the oviraptorosaurian clade Oviraptoridae (other members of Oviraptorosauria include the caenagnathids, *Avimimus*, *Caudipteryx* and *Microvenator*). It has been proposed that Oviraptoridae consists of two radiations: the mostly crested oviraptorines, and the crest-less ingeniines, though some recently described taxa do not fit neatly into this dichotomy. Starting with the oviraptorines, we begin with *Oviraptor philoceratops*, the first of them to be named (Osborn 1924). Ironically, it's not that well known and the fact that, until recently, most Mongolian oviraptorids were assumed to be referable to this genus means that very few illustrations of *Oviraptor* really do depict this genus. It turns out that *Oviraptor* was comparatively long-skulled for an oviraptorid and it may have been the most basal oviraptorine (Clark *et al.* 2002). Some kind of premaxillary crest was present, but the skull is not well preserved enough to determine its original shape.

Fig. 37. A scene depicting Late Cretaceous life from the Gobi Desert of Mongolia (this illustration was produced in 1993 and depicts some of the animals inaccurately; they are also not correctly scaled, and they did not all live in the same time and place!). The oviraptorine GIN 100/42 and the ingeniine *Conchoraptor* are in the background, the enantiornithine *Gobipteryx* flies overhead, and (in the foreground) the alvarezsaurid *Mononykus* defends its chicks from the monstersaurian lizard *Estesia*.

Long labelled as a new species of *Oviraptor* is the tall-crested animal now known as *Rinchenia mongoliensis*. The generic name has been kicking around for a while but its formal use in a major compendium (Osmólska *et al.* 2004) means that it is now 'officially' in use. In *Rinchenia* the crest is mound-shaped and extends for most of the length of the skull. There is another specimen often illustrated as an *Oviraptor* – in fact normally labelled as *Oviraptor philoceratops* (e.g., Barsbold *et al.* 1990) – and this is GIN 100/42 (Fig. 37). It's short-skulled and sports a rounded crest decorated with lateral accessory openings. Though regarded by some as a new species of *Citipati* (on which see below), one unpublished study found it to be closest to

O. philoceratops. Whatever, "because [GIN 100/42] is much better preserved than the holotype of *O. philoceratops* it has often been relied upon for anatomical details of this species, so caution should be used in referring to previous characterization of this species" (Clark *et al.* 2002, p. 21). It's a new taxon, whatever its affinities.

Finally among the oviraptorines, there is *Citipati osmolskae*. This was a large animal compared to most other oviraptorids with a distinctive rostrodorsally inclined back part to the skull, a large, sub-oval naris, and a premaxillary crest that is continuous with the premaxilla's rostral margin (Clark *et al.* 2001). The famous 'big momma' specimen (an animal preserved atop an egg-filled nest), and an embryonic specimen, also appear referable to *Citipati*. The embryo is significant in that it was the specimen that demonstrated that most/some eggs attributed to *Protoceratops* were actually oviraptorid eggs. I don't have time to cover that story in depth now, and it's reasonably well known now anyway. Not all studies agree that *Citipati* is an oviraptorine – it may instead be an ingeniine.

We move now to the ingeniines, a group separated from oviraptorines by Rinchen Barsbold, mostly due to differences in hand structure (Barsbold *et al.* 1990). Barsbold later thought that ingeniines were distinct enough to deserve their own 'family', Ingeniidae, though this has not been supported by more recent studies. The type taxon of the group is '*Ingenia*' *yanshini*, a small, robust species ('small' = c. 1.5 m long) with particularly short arms, a massive thumb, and robust, short feet. It has recently been discovered that the name '*Ingenia*' is preoccupied, and this explains the quotes used here. A replacement name will appear at some stage, but hasn't done so yet. Rather similar to '*Ingenia*', though more gracile, is *Conchoraptor gracilis*. The short, rounded skull of this species has been illustrated and labelled several times as representing *Oviraptor*, though it is clearly distinct from that taxon. It has a distinctive subvertical, dorsoventrally elongate naris. *Khaan mckennai*, described for two outstanding complete skeletons, is similar to *Conchoraptor* but has a more horizontally-aligned naris and more strongly curved manual unguals (Clark *et al.* 2001). If I remember correctly, the ICZN prefers it if 'Mc' names are converted to 'mac' spellings when used in binomials. If this is right, then the specific name here should really have been 'mackennai'. As an example, the ichthyosaur *Macgowania* was named for Chris McGowan.

Also probably part of Ingeniinae is *Heyuannia huangi*, a taxon named in 2002 and known from at least four individuals, one of which is an almost complete skeleton (Lü 2002, Lü *et al.* 2005). It was also small (less than 2 m long), crest-less, with a longer neck and shorter back than other taxa, and a particularly long ilium.

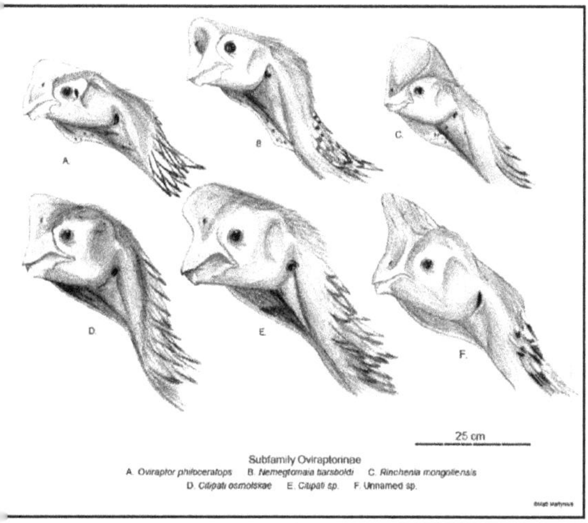

Fig. 38. Oviraptorine heads compared. Illustration by Matt Martyniuk.

The most recently named oviraptorid is *Nemegtomaia barsboldi* from the Maastrichtian Nemegt Formation (Lü *et al.* 2004, 2005). Lü *et al.* (2004) originally named this taxon *Nemegtia*, but that turned out to be preoccupied by an ostracod from the Nemegt Formation (shades of '*Ingenia*'). *Nemegtomaia* is reasonably well known, with a near-complete skull. Prior to this, the specimen had been figured, but identified (without good reason) as a new specimen of '*Ingenia*' (Lü *et al.* 2002). However, unlike '*Ingenia*', *Nemegtomaia* is an oviraptorine, and it's probably closely related to *Citipati*.

We also have *Shixinggia oblita* from the Maastrichtian Pingling Formation of Guangdong Province, China, and known from pelvic, hindlimb and vertebral material (Lü & Zhang 2005). The affinities of this taxon are uncertain – it may be a caenagnathid and not an oviraptorid. For various reasons I'm very interested right now in postcranial pneumaticity, and for this reason I'll say that *Shixinggia* (and *Heyuannia*) are particularly interesting.

Also on the subject of caenagnathids, there is *Hagryphus giganteus*, just described from the Campanian Kaiparowits Formation of Utah (Zanno & Sampson 2005). Though only known from a hand, a fragmentary distal part of a radius and some fragments from the foot, it's clearly distinct from other North American taxa. It was large – bigger even than *Chirostenotes* (and 'large' here means >3 m long). And this isn't the end of it as at least a few other taxa (mostly oviraptorids) have been figured in the literature and appear to represent valid taxa, but have yet to be named or described.

The expanding taxonomic sample and diversity of these animals is the exciting area that I've covered here, but there's so much other interesting stuff that could be said about oviraptorosaurs. There's what we know of their reproductive behaviour (gleaned from nests with eggs and even shelled eggs discovered inside a pelvis), the affinities and anatomy of the basal forms from Lower Cretaceous China (*Caudipterx*, *Protarchaeopteryx* and *Incisivosaurus*), the controversy over their diet and ecology, their controversial phylogenetic position (some workers maintain that they are flightless birds: let me take this opportunity to point out that analysis of the data indicates that this view is erroneous, and that these theropods are outside of the clade that includes archaeopterygids and modern birds), and the controversy over their alleged presence in the English Lower Cretaceous (Naish & Martill 2002).

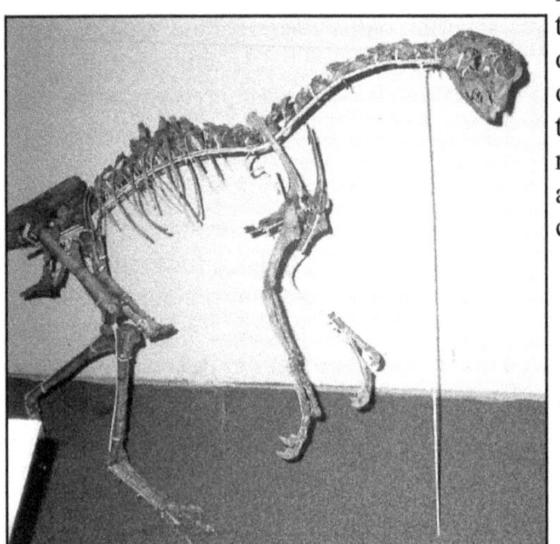

Fig. 39. Skeleton of *Citipati*, an oviraptorine oviraptorid. Oviraptorids might have superficially resembled flightless birds when alive, but many details of their behaviour and biology remain enigmatic.

Refs - -

- Barsbold, R., Maryańska, T. & Osmólska, H. 1990. Oviraptorosauria. In Weishampel, D. B., Dodson, P. & Osmólska, H. (eds) *The Dinosauria*. University of California Press (Berkeley), pp. 249-258.
- Clark, J. M., Norell, M. A. & Barsbold, R. 2001. Two new oviraptorids (Theropoda: Oviraptorosauria), Upper Cretaceous Djadokhta Formation, Ukhaa Tolgod, Mongolia. *Journal of Vertebrate Paleontology* 21, 209-213.
- Clark, J. M., Norell, M. A. & Rowe, T. 2002. Cranial anatomy of *Citipati osmolskae* (Theropoda, Oviraptorosauria), and a reinterpretation of the holotype of *Oviraptor philoceratops*. *American Museum Novitates* 3364, 1-24.
- Gee, H. & Rey, L. V. 2003. *A Field Guide to Dinosaurs*. Quarto Publishing (London).
- Lü, J. 2002. A new oviraptorosaurid (Theropoda: Oviraptorosauria) from the Late Cretaceous of southern China. *Journal of Vertebrate Paleontology* 22, 871-875.
- Lü, J., Dong, Z., Azuma, Y., Barsbold, R. & Tomida, Y. 2002. Oviraptorosaurs compared to birds. In Zhou, Z. & Zhang, F. (eds). *Proceedings of the 5th Symposium of the Society of Avian Paleontology and Evolution*. Science Press (Beijing), pp. 175-189.
- Lü, J., Tomida, Y., Azuma, Y., Dong, Z. & Lee, Y.-N. 2004. New oviraptorid dinosaur (Dinosauria: Oviraptorosauria) from the Nemegt Formation of southwestern Mongolia. *Bulletin of the National Science Museum, Tokyo, Series C* 30, 95-130.
- Lü, J., Tomida, Y., Azuma, Y., Dong, Z. & Lee, Y.-N. 2005. *Nemegtomaia* gen. nov., a replacement name for the oviraptorosaurian dinosaur Nemegtia Lü et al., 2004, a preoccupied name. *Bulletin of the National Science Museum, Tokyo, Series C* 31, 51.
- Lü, J. & Zhang, B.-K. 2005. A new oviraptorid (Theropod [sic]: Oviraptorosauria) from the Upper Cretaceous of the Nanxiong Basin, Guangdong Province of southern China. *Acta Palaeontologica Sinica* 44, 412-422.
- Naish, D. & Martill, D. M. 2002. A reappraisal of *Thecocoelurus daviesi* (Dinosauria: Theropoda) from the Early Cretaceous of the Isle of Wight. *Proceedings of the Geologists' Association* 113, 23-30.
- Osborn, H. F. 1924. Three new Theropoda, Protoceratops Zone, central Mongolia. *American Museum Novitates* 144, 1-12.
- Osmólska, H., Currie, P. J. & Barsbold, R. 2004. Oviraptorosauria. In Weishampel, D. B., Dodson, P. & Osmólska, H. (eds) *The Dinosauria, Second Edition*. University of California Press (Berkeley), pp. 165-183
- Zanno, L. E. & Sampson, S. D. 2005. A new oviraptorosaur (Theropoda, Maniraptora) from the Late Cretaceous (Campanian) of Utah. *Journal of Vertebrate Paleontology* 25, 897-904

Chapter 12:

Lots of sauropods, or just a few sauropods, or lots of sauropods?

As discussed in the previous chapter (remember: this text was originally written in February 2006), early last week Mike P. Taylor (Fig. 41) and I spent the better part of a day working on Wealden sauropods in the Natural History Museum collections. The main point of this excursion was to have our final look ('final' as in last look before we submit our paper) at a specimen that we've been working on: a highly unusual, in fact deeply weird, new taxon collected in the late 1800s from the Ashdown Beds Formation (Valanginian) near Hastings. I won't be saying more about it until it's published, but its 'discovery' has prompted me to revisit the problematic area of Wealden sauropod diversity. Well, ok, I'm 'revisiting' that area anyway right now, what with the British dinosaurs manuscript, but still. Was sauropod diversity in the Early Cretaceous of England high, or low?

Fig. 40. Representatives of some of the sauropod groups present in the sediments of the English Wealden Supergroup. From top to bottom: a diplodocid diplodocoid, a titanosaur, and a brachiosaur.

First things first: it's misleading to think of the 'Wealden' as a short chunk of time, as this term actually applies to three formal stratigraphic groups (the Berriasian-Barremian Hastings Beds Group and Weald Clay Group of the mainland, and the Hauterivian-Aptian Wealden Group of the Isle of Wight). 'Berriasian to Aptian' is something like 30 million years, so that's a lot of time for a whole lot of species to come and go. Whole assemblages of species in fact. Certainly the animals known from the Hastings Beds Group are not the same as those of the Barremian Wessex Formation (the best known dinosaur-bearing unit in the Wealden Group). The Hastings Beds Group has yielded *Pelorosaurus conybeari* (which is actually the same thing as *Cetiosaurus brevis*, but we won't go there right now) and *P. becklesii* (which clearly isn't congeneric with *P. conybeari*). *P. conybeari*, based on caudal vertebrae, chevrons and a large and gracile humerus, is a basal titanosauriform, perhaps a brachiosaurid. *P. becklesii*, named for tremendously short and robust forelimb elements and skin, is clearly a titanosaur. This is significant as it's among the oldest of verified body fossil representing this clade. However, there are trackways that appear to have been produced by titanosaurs from the Bathonian of Ardley, Oxfordshire (Day *et al.* 2002, 2004), so titanosaurs had been around since the Middle Jurassic at least.

That's not all – the Hastings Beds Group has also yielded a single metacarpal which has been identified as belonging to a diplodocid: that's right, not just a diplodocoid, but a diplodocid. Angela Milner initially made this identification, and more recently I've had it verified by Matt Bonnan (pers. comm. 2006). Then we have the new taxon that Mike and I are looking it. It doesn't seem that any of these animals can be conspecific, so, we have: a basal titanosauriform (and possible brachiosaurid), a good titanosaur, a diplodocid, and a new taxon that isn't any of these.

What about the Wessex Formation, where 30-odd Barremian dinosaur species are known? Naish & Martill (2001) and Naish (2005) thought that sauropod diversity here was quite high, with a possible camarasaurid (*Chondrosteosaurus*), a few brachiosaurids, a titanosaur (*Iuticosaurus valdensis*), and an unnamed diplodocoid (represented by a chevron as well as isolated teeth and other elements). This now seems erroneous, or perhaps over-optimistic. The characters supposedly indicating camarasaurid status for *Chondrosteosaurus* are rubbish, and it's more likely a basal titanosauriform (and it's non-diagnostic anyway). The supposed diplodocoid bits are controversial, and the 'sled-like' morphology initially used to support a diplodocoid identity for the famous chevron (Charig 1980) is now thought to have been primitive for neosauropods, rather than derived for diplodocoids (see Upchurch 1998). It's also difficult to be sure that there's more than one brachiosaurid. The evidence for titanosaurs (procoelous caudal vertebrae, with a distinctively located neural arch) is reasonable, but the remains aren't diagnostic.

Accordingly, when Paul Upchurch spoke about Wessex Formation sauropod diversity at the British dinosaurs Palaeontological Association Review Seminar in November 2003 (co-hosted by Dinosaur Isle Museum and the University of Portsmouth) he cautioned that diversity might have been over-estimated, and that titanosauriforms accounted for what diversity there was.

But while some groups – like camarasaurids – have now been removed from the list of taxa, new ones have been added (as it happens, camarasaurids were apparently present elsewhere in Lower Cretaceous Europe, but more about that another time). Firstly, there are definitely brachiosaurids present in the Wessex Formation: the big MIWG 7306 cervical vertebra that I and colleagues described in 2004 (Naish *et al.* 2004) is clearly more like the vertebrae of *Brachiosaurus* and *Sauroposeidon* than anything else, and some of the dorsal vertebrae referred by Blows (1995) to his problematic taxon *Eucamerotus foxi* are also clearly brachiosaurid in the strict sense. Then there are the clearly titanosaurian caudal vertebrae, such as those described by Le Loeuff (1993). But there's more.

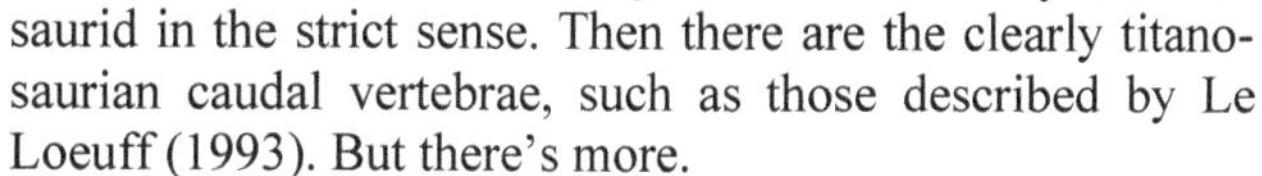

Fig. 41. Partner-in-crime Michael P. Taylor, posing in a museum basement with a Wealden sauropod vertebra.

A big and highly distinctive, pristine tooth forms the holotype of *Oplosaurus armatus*. Naish & Martill (2001), the idiots, thought that this might be another brachiosaurid, but that's clearly wrong: Canudo *et al.* (2002) suggested that *Oplosaurus* might instead have been closer to *Camarasaurus*, and Upchurch *et al.* (2004) regarded it as a distinct neosauropod of uncertain affinities. Then there are the brachiosaurid teeth named *Pleurocoelus valdensis*: most authors have regarded these as non-diagnostic bits of brachiosaurid, but Ruiz-Omeñaca & Canudo (2005) have argued that they're diagnostic and really represent a good species. Whatever they are, they're clearly something else from *Oplosaurus*.

And then, finally, there are the diplodocoid teeth figured by Naish & Martill (2001) and since identified as.... rebbachisaurid teeth! The recognition of these teeth as rebbachisaurid verifies the presence of diplodocoids in the Wessex Formation. Ok, so they aren't diplodocid-like forms as Charig and others thought when they were looking at the chevron, but they're diplodocoids nonetheless. In some senses this discovery isn't a surprise as rebbachisaurids have lately been reported from Barremian-Aptian Spain (Fernández-Baldor *et al.* 2001, Pereda Suberbiola *et al.* 2003), but it now means that the oldest rebbachisaurids in the world are from England. Huh, so much for this being an exclusively Gondwanan clade.

So, in the Wessex Formation we have... an unusual neosauropod of uncertain affinities (*Oplosaurus*), definite brachiosaurids, definite titanosaurs, and a rebbachisaurid. That's not bad in terms of diversity. To remind you, in the older Hastings Beds Group we have a basal titanosauriform (and possible brachiosaurid), a titanosaur, a diplodocid, and a new and unusual taxon that isn't any of these. So, also, not bad in terms of diversity, and in fact not that different from the Wessex Formation roster, which implies stability from the point of view of long-term diversity.

Refs - -

- Blows, W. T. 1995. The Early Cretaceous sauropod dinosaurs *Ornithopsis* and *Eucamerotus* from the Isle of Wight, England. *Palaeontology* 38,187-197.
- Canudo, J. I., Ruiz-Omeñaca, J. I., Barco, J. L. & Royo Torres, R. 2002. ¿Saurópodos asiáticos en el Barremiense inferior (Cretácico Inferior) de España? *Ameghiniana* 39, 443-452.
- Charig, A. J. 1980. A diplodocid sauropod from the Lower Cretaceous of England. In Jacobs, L. L. (ed.) *Aspects of Vertebrate History, Essays in honour of E. H. Colbert.* Museum of Northern Arizona Press, pp. 231-244.
- Day, J. J., Norman, D. B., Gale, A. S., Upchurch, P. & Powell, H. P. 2004. A Middle Jurassic dinosaur trackway site from Oxfordshire, UK. *Palaeontology* 47, 319-348.
- Day, J. J., Upchurch, P., Norman, D. B., Gale, A. S. & Powell, H. P. 2002. Sauropod trackways, evolution, and behavior. *Science* 296, 1659.
- Fernández-Baldor, T., Pereda Suberiola, X., Huerta Hutado, P., Izquierdo, L. A., Montero, D. & Pérez, G. 2001. Descripción preliminar de un dinosaurio rebaquisáurido (Sauropoda Diplodocoidea) del Cretácico Inferior de Burgos (España). *II Jornadas de Paleontología de Dinosaurios y su Entorno.* Salasa de los Infantes (Burgos, España), 203-211.
- Le Loeuff, J. 1993. European titanosaurs. *Revue de Paléobiologie* 7, 105-117.
- Naish, D. 2005. The sauropod dinosaurs of the Wealden succession (Lower Cretaceous) of southern England. *The Quarterly Journal of the Dinosaur Society* 4 (3), 8-11.
- Naish, D. & Martill, D. M. 2001. Saurischian dinosaurs 1: Sauropods. In Martill, D. M. & Naish, D. (eds) *Dinosaurs of the Isle of Wight.* The Palaeontological Association (London), pp. 185-241.
- Naish, D., Martill, D. M., Cooper, D. & Stevens, K. A. 2004. Europe's largest dinosaur? A giant brachiosaurid cervical vertebra from the Wessex Formation (Early Cretaceous) of southern England. *Cretaceous Research* 25, 787-795.
- Pereda Suberbiola, X., Torcida, F., Izquierdo, L. A., Huerta, P., Montero, D. & Perez, G. 2003. First rebbachisaurid dinosaur (Sauropoda, Diplodocoidea) from the early Cretaceous of Spain : palaeobiogeographical implications. *Bulletin de la Societe Geologique de France* 174, 471-479.
- Ruiz-Omeñaca, J. I. & Canudo, J. I. 2005. "*Pleurocoelus*" *valdensis* Lydekker, 1889 (Saurischia, Sauropoda) en el Cretácico Inferior (Barremiense) de la Península Ibérica. *Geogaceta* 38, 43-46.
- Upchurch, P. 1998. The phylogenetic relationships of sauropod dinosaurs. *Zoological Journal of the Linnean Society* 124, 43-103.
- Upchurch, P., Barrett, P. M. & Dodson, P. 2004. Sauropoda. In Weishampel, D. B., Dodson, P. & Osmólska, H. (eds) *The Dinosauria, Second Edition.* University of California Press (Berkeley), pp. 259-322.

Chapter 13:
No no no no NO: the Herring gull is NOT a ring species!

So many animals that we totally take for granted are actually, when you think about them, really remarkable. In keeping with the theme of alien invaders (see Chapters 8 and 10), I was going to talk about Collared doves *Streptopelia decaocto*, and the various egret species that are presently 'invading' the British Isles. But it's another ordinary, yet remarkable, bird that I'm going to talk about now, and it's the Herring gull *Larus argentatus* (Fig. 42), a behaviourally flexible, adaptable and widespread bird that inhabits Eurasia and North America. It's a large bird that can exceed 70 cm in length and have a wingspan of 1.3 m. It's also a supreme generalist, capable of thriving on all kinds of food, and this explains its highly successful colonisation of urban environments.

Fig. 42. Herring gull. Photo by Neil Phillips.

While, as I said, we take urban gulls for granted, if you think about it, it's pretty amazing that a seabird with a 1.3 m wingspan has successfully colonised towns and cities. They've learnt to drop shelled prey from heights in order to break it open (there's the famous anecdote of a New Jersey highway bridge littered with clam shells dropped by enterprising gulls) and they also like to wash their food in rock-pools. Their breeding behaviour and how they and their chicks respond to stimuli is tremendously well studied, with the studies of Goethe, Tinbergen and others being classic, pioneering works in ethology.

But perhaps what makes the Herring gull most 'famous' among biologists is its alleged status as the examplar *par excellence* of a 'ring species'. As we'll see below, this concept has now been all but debunked, and it's because of this that I'm surprised whenever I see continuing references to it. Who might be the most naughty of recent offenders, I hear you ask? Richard Dawkins. Yes, he of *Selfish Gene* fame. I noted a while back that I was reading *The Ancestors Tale* (Dawkins 2004). It's a good book, sure, but I've found it a tedious read (with awful artwork), and to be honest I've repeatedly given up on it and moved on to Mayor's *The First Fossil Hunters*, Pianka & Vitt's *Lizards: Windows to the Evolution of Diversity*, Patterson's *The Lions of Tsavo* and Fisher & Lockley's *Seabirds*.

The ring species concept is a big deal as it proposes a mechanism for one of the most important questions of evolutionary biology: the origin of species. And it was study of Herring gulls that led the late great Ernst Mayr (1942) to argue that speciation in Herring gulls had occurred by way of 'isolation by distance': while adjacent populations would be able to breed with one another, the genetic distance resulting from the expansion of a species far from its centre of origin would eventually produce an 'end' population so far removed from its ancestor that it would be incapable of interbreeding with it. It would now be a separate species. Mayr proposed that exactly this has happened, and that, after originating in the Aralo-Caspian region, Herring gulls had moved north to the Arctic Ocean. Here they expanded west, giving rise to dark-mantled forms of the Lesser black-backed gull (Fig. 43), and also east, giving rise to the pale-mantled forms of Herring gull of Siberia and North America. Finally, North American Herring gulls crossed the Atlantic to invade Europe, and here they encountered the Lesser black-backed gulls that marked the other end of the ring. Both end points had now reached reproductive isolation, and today coexist as distinct species.

Fig. 43. Lesser black-backed gull.
Photo by Neil Phillips.

Don't get me wrong: there's good evidence that speciation does occur in this way in some instances (e.g., in Californian *Ensatina* salamanders, and southern Asian leaf warblers), but it seems to be very rare. And, as it happens, new study indicates that it did not happen in the case of Herring and Lesser black-backed gulls. Firstly, the taxa regarded by Mayr as subspecies of these two are now regarded as distinct enough to be regarded as separate species. The Aralo-

Caspian gull regarded by Mayr (1942) as the ancestral Herring gull population is the Caspian or Steppe gull *L. cachinnans*, and the Mediterranean and eastern Atlantic gull thought by Mayr (1942) to be a westward excursion of the Caspian gull is the Yellow-legged gull *L. michahellis*. Furthermore, if a recently proposed subdivision of Lesser black-backed gulls is accepted (Sangster *et al.* 1998), then the proper name for this species is *L. graellsii*, and two taxa previously ranked as subspecies – the Tundra gull *L. heuglini* and Baltic gull *L. fuscus* – should be separated as species. Of course you could bring in here the debate over the whole subspecies concept: there are over 20 of these in the Herring gull-Lesser black-backed gull complex, so... gack... how many species should we be recognising? And once we do start to recognise at least some of these taxa as species, doesn't this negate the whole *raison d'être* of Mayr's proposed ring?

Fig. 44. Great black-backed gull: the largest gull, its wingspan may reach 1.7 m and it can weigh 2 kg. Photo by Burkhard Plache.

Furthermore, in a study of mtDNA in white-headed gulls, Liebers *et al.* (2004) found that white-headed gull phylogeny and biogeography was far more complex than Mayr and others had thought. Yellow-legged gulls were not closest to Caspian gulls, but instead seem to have descended from a North Atlantic ancestral population. A separate ancestral population moved north from the Aralo-Caspian region toward the British Isles, giving rise to the Lesser black-backed gull, and east toward Siberia and North America, where Tundra gulls, Slaty-backed gull *L. schistisagus* and Glaucous-winged gulls *L. glaucescens* arose. Intriguingly, the Great black-backed gull *L. marinus* (Fig. 44) was not an outgroup to the Herring gull-Lesser black-backed complex as usually thought, but was actually nested within the complex and probably evolved (in allopatry with *L. argentatus*) in northeastern N. America. Glaucous gulls *L. hyperboreus* (Fig. 45) and Kelp gulls *L. dominicanus* were also nested within *L. argentatus*.

The Kelp gull is unique to the Southern Hemisphere, so Liebers *et al.* (2004) concluded that it must have evolved via long-distance colonisation "from the same ancestral population as the Lesser black-backed gull, suggesting that its ancestors were highly migratory, as nominate Lesser black-backed gulls still are today" (p. 895). The central Asian *L. mongolicus* didn't originate from Caspian gulls, but from Pacific gulls close to *L. schistisagus*.

Fig. 45. Glaucous gull. Photo by Alastair Rae.

If all of this seems horribly confusing, I think that's because it is. The number and variety of white-headed gull taxa is baffling and sorting out any kind of historical pattern is highly, highly difficult. The picture is made more complex by the fact that populations which appear to belong to different lineages (e.g., *L. michahellis* from the Atlantic Iberian coast and western European *L. argentatus*) look similar, apparently due to convergence (Pons *et al.* 2004). It's also difficult to tell whether strong genetic similarities reported between some taxa – such as Baltic gulls and Tundra gulls for example – result from recent separation or from ongoing gene flow (Liebers & Helbig 2002). There's also the interesting discovery that supposed hybrids (of *L. hyperboreus* and *L. argentatus*) turned out to be light-winged *L. argentatus* founders that were expanding their range (Snell 1991).

But, most importantly, support for the simple ring model is lacking as there is no evidence that North American Herring gulls recolonised Europe to encounter the Lesser black-backed gulls that marked the other end of the ring. However, the great irony of all this is that *L. graellsii* is presently spreading westwards, and may eventually colonise North America. Should it do this (right now it breeds as far west as Greenland), it will encounter the North American Herring gull *L. argentatus smithsonianus*, and if these two forms prove incapable of interbreeding, then the ring species model would have been fulfilled… albeit it by birds moving from east to west, rather than west to east as Mayr proposed.

Well, all of that was pretty complicated. Feel free not to remember it, but take home at least the title of Liebers *et al.*'s paper: 'The herring gull complex is not a ring species'.

Refs - -

- Dawkins, R. 2004. *The Ancestor's Tale*. Weidenfeld & Nicolson (London).
- Liebers, D., de Knijff, P. & Helbig, A. J. 2004. The herring gull complex is not a ring species. *Proceedings of the Royal Society of London B* 271, 893-901.
- Liebers, D. & Helbig, A. J. 2002. Phylogeography and colonization history of Lesser black-backed gulls (*Larus fuscus*) as revealed by mtDNA sequences. *Journal of Evolutionary Biology* 15, 1021-1033.
- Mayr, E. 1942. *Systematics and the Origin of Species*. Columbia University Press (New York).
- Pons, J.-M., Crochet, P.-A., Thery, M. & Bermejo, A. 2004. Geographical variation in the yellow-legged gull: introgression or convergence from the herring gull? *Journal of Zoological Systematics & Evolutionary Research* 42, 245-256.
- Sangster, G., Hazevoet, C. J., Berg, A. & van den Roselaar, C. S. 1998. Dutch avifaunal list: species concepts, taxonomic stability, and taxonomic changes in 1998. *Dutch Birding* 20, 22-32.
- Snell, R. R. 1991. Variably plumaged Icelandic Herring gulls reflect founders not hybrids. *The Auk* 108, 329-341.

Chapter 14:

Cryptic dinosaur diversity, 'real taxon' counts, curse of the *nomina dubia*, and the holy grail: matrix-assisted laser desorption ionization

The unthinkable has happened. After more than three months of work, the 20,000-word, 100-page review manuscript on British dinosaurs has finally been submitted (by the way, 20,000 words is horribly close to the 30,000 words required as minimum for a phd science thesis). Well, actually, it hasn't been submitted, but it's now out of my hands and will be submitted within the next few days, so for the purposes of this text let's just pretend that it has been submitted. I learnt a lot from the project, and it also got me thinking about some issues that extend beyond the subject of British dinosaurs.

Here's what's bothering me. Of 108 named British dinosaur taxa (representing the valid species and the *nomina dubia*), over 50% are *nomina dubia* (that is, based on remains that lack autapomorphies [unique, diagnostic characters]). It was only natural that in the discussion section, we (= Naish & Martill) made a few comments on British dinosaur diversity: on how the taxonomic spread compares with global dinosaur diversity, and on how many species we have for each group. The contention we hold (in the present version of the MS: this might change after review) is that the 'real diversity' preserved in the fossil record is higher than the diversity count indicated only by the valid taxa. How so?

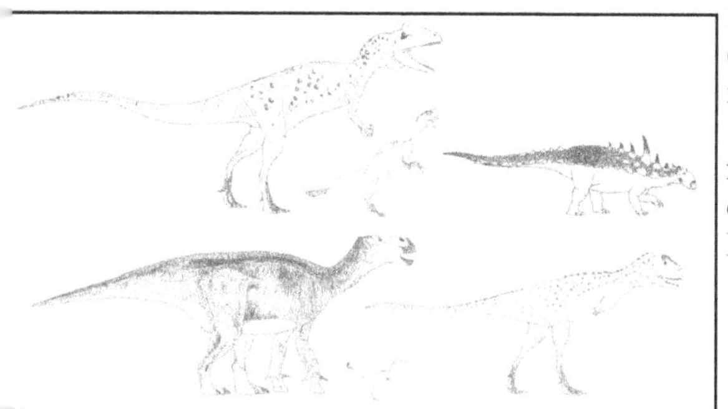

Fig. 46. An entirely random and pointless assortment of dinosaurs.

Why valid species aren't the only species we should count

Consider that, rather than being useless, indeterminate objects that do not really reflect biological entities, some *nomina dubia* surely represent real taxa, it's just that we lack the data to diagnose them by autapomorphies. Many *nomina dubia* come from geological units where there are no named valid taxa. While a *nomen dubium* might lack unique characters, if it's not definitely referable to anything else, then surely it really is a valid taxon? There's already a precedent for recognising taxa that lack autapomorphies: they're called metataxa, and I think that at least some *nomina dubia* should be regarded as such. As soon as they are, the 'real taxon' count goes up. In the case of British dinosaurs, it does so by 20%.

Then there are unnamed taxa: specimens that can be identified to a clade but remain unnamed because, like *nomina dubia*, they lack autapomorphies. In many cases, however, they represent the only member of their group known from the relevant unit. In the British record, examples of such are the Bexhill diplodocid, Wessex Formation rebbachisaurid (see Chapter 12) and a large theropod from Stonesfield that is not the same thing as *Megalosaurus*. While such specimens shouldn't be given binomials (pending the discovery of better material), they should be regarded as countable taxa. Inclusion of them also therefore ups the 'real taxon' count, and in the case of British dinosaurs, that count now increases by a further c. 22%.

These observations mean that diversity counts for dinosaurs (and other fossil groups) are necessarily under-estimates, as they only account for named valid species. But, if anything, dinosaur diversity studies have been ultra-conservative, and have not only ignored this cryptic diversity, they have also downplayed even the counts of named valid taxa. In a recent talk on dinosaur diversity, Mike Benton (2005) argued that dinosaurs were more taxonomically inflated than most other animal groups. Across living and fossil taxa, it has been shown that about 20% of all 'taxa' eventually turn out to be synonymous with others (Alroy 2002), but for dinosaurs Mike argued that the figure was more like 50%. So while dinosaur workers are patting themselves on their backs for being in a golden age of taxonomic discovery (the number of named dinosaurs has increased 70% since 1990), Mike's contention is that most of this is over-zealous and/or over-enthusiastic. But, hold on, exactly who says that as many as 50% of dinosaur taxa are junior synonyms? I'm not going to name names here, or go into this in any depth (it would involve detailed discussion of specific groups and their taxonomic histories), but I am not convinced that this view is valid.

Fig. 47. Dinosaurs like *Iguanodon* (in the classic, highly inclusive, polytypic sense) encompass a huge amount of diversity and probably represent 'over-lumping'.

'Un-lumping' in the extant tetrapod fauna, and the recognition of cryptic diversity

It is very easy to be critical of those systematists that we term 'splitters': those who erect new species and/or genera based on the smallest of differences. Mostly the criticisms are justified, as true splitters really do use the tiniest, tiniest little differences to erect their new species/genera, and subsequent work usually shows that these 'species'/'genera' fall within the range of individual, ontogenetic or sexual variation of a previously named taxon. Fair enough, I've collapsed over-split taxa myself (Martill & Naish, in press [on azhdarchoid pterosaurs]). But there is less criticism of those systematists termed 'lumpers': those who synonymise species and/or genera and are prepared to accept considerable variation within their concept of species and genera. And it seems that, with dinosaurs, the lumping may actually be more laissez-faire than the splitting: in the relevant cases, the workers concerned have synonymised taxa because of gestalt similarities. NOT because detailed, careful work has shown that the taxa really can be regarded as conspecific or congeneric.

If dinosaurs really do get lumped more than they already are, it strikes me that the trend within dinosaur systematics is in direct opposition to that occurring among extant tetrapods. No, the number of recognised taxa is not going down because of increasing rates of synonymy, it is going up as specialists, working across various tetrapod groups, are resurrecting taxa from synonymy, or are discovering cryptic diversity and thereby coining new species and genera. On the first point (resurrection of taxa from synonymy), it seems that the 20th century trend of synonymising the many taxa erected by Victorian biologists was often more laissez-faire and over-zealous than was the splitting that occurred before. Indeed the standard taxonomic compilations that exist for extant tetrapods all now seem guilty of lumping on a massive, uncritical scale. Judging by statements made in the literature, this appears to be the case for big mammals and birds more than for other groups.

It's easy to observe this in action on bird systematics: most specialists agree that certain 'standard' reference works (e.g., Peters 1951) were just too heavy-handed with their lumping. Judging by the 'un-lumping' that's gone on so far, there is every indication that the count of c. 9000 extant bird species usually given in the literature is an under-estimate.

On the second point (the discovery of cryptic diversity), DNA-based studies on extant species seem – as a generalisation – to mostly discover that groups include more diversity than was thought before, not less. For a few good, specific examples of this sort of thing, see Wüster & Thorpe (1994), Mayer & von Helversen (2001), Glaw *et al.* (2001), Glaw & Vences (2002a, b), Vallan *et al.* (2003), Parra-Olea *et al.* (2005) and Olsson *et al.* (2005). Hand-in-hand with this is the debate over the status of 'subspecies', a complex issue that I'll avoid here. I agree with those who argue that the subspecies concept is not useful, and that definable taxonomic units are the same thing as species (Zink 2004). Again, when such taxonomic units are transformed into species, diversity counts go up. For specific examples see Collins (1991) [where 55 North American amphibians and reptiles were raised to species rank]; Sibley & Monroe (1991) [where the number of extant bird species was upped from c. 9000 to 9672]; Robertson & Nunn (1998) [where 10 albatross taxa were raised to species rank]; and Alström *et al.*

(2003) [where 16 wagtail taxa are noted as 'new' species if the phylogenetic species concept is applied].

On birds again, one study has even claimed that the recognition of all of this cryptic diversity "would lead to a doubling of the currently recognized species diversity of birds" such that there might be 20,000 extant species (Zink 1996). Indeed it is widely acknowledged by those who have criticised both lumping, and the subspecies concept, that lumping is potentially harmful both to diversity studies and to conservation. It essentially downplays true potential diversity and diverts interest, research and conservation concern, plus most subspecies were named for convenience anyway and don't reflect real evolutionary units (Zink 2004).

In view of both the discovery of cryptic diversity and the transformation of 'subspecies', and keeping in mind that a significant percentage of extant species are all but indistinguishable osteologically, it is a safe assumption that any measure of diversity among fossil species is an under-estimate. Ralph Molnar (1990) made this point some time ago, but I've yet to see a study on dinosaur diversity that acknowledges or refers to his point. I realise that I'm on dangerous ground in employing this argument, as it is simply not possible to examine fossil taxa at the same level of detail. The new data on extant diversity mostly comes from genetics, and there isn't too much data of that sort available when we're thinking of Mesozoic animals. So this is sort of an apples and oranges situation, and claims that 'fossil species and/or genera are over-lumped' come down to nothing more than arm-wavy assertions or speculations. Given confounds like ontogeny and individual variation, and then the really really cool stuff like phenotypic plasticity and resource polymorphism, any claims that fossil taxa are, or are not, synonymous are by nature subjective, and actually unknowable. Right?

A holy grail for palaeodiversity?

So wouldn't it be nice if there were some sort of holy grail. Some magical method, akin to mtDNA analysis, that might allow us to work out whether two fossil specimens were or were not members of the same species or genus? Well here's the big news: there is. Sort of.

Osteocalcin is a mineral-associated protein, exclusive to vertebrates, that occurs within bone matrix. It is in fact the second most abundant protein within bone, and only a tiny amount of bone is needed to extract it. After extraction and purification, the protein is sequenced with matrix-assisted laser desorption ionisation mass spectrometry (MALDI-MS). Here's the big news for the

Fig. 48. Are extremely similar fossil tetrapods – like *Tarbosaurus* and *Tyrannosaurus* shown here – members of the same genus or not? Experts disagree. It would be great if there were some way to know.

world of palaeontology: (1) osteocalcin sequences appear to be genus-specific, and (2) osteocalcin remains stable and detectable in old fossil bone and, while to date it's only been applied to Pleistocene fossils, no-one yet knows how far back it can be detected. Presumably this is because no-one has yet really looked (please tell me if you know otherwise). I have been reliably informed by a geochemist than there is no theoretical reason why osteocalcin can't be detected in bone that is tens or millions, or hundred of millions, of years old.

To date two studies have applied MALDI-MS to fossil samples, and both produced positive, highly encouraging results. Nielsen-Marsh *et al.* (2002) looked at fossil bison sequences, and Nielsen-Marsh *et al.* (2005) looked at sequences in Neanderthals. In both cases, the fossil taxon proved to have osteocalcin sequences that were, firstly, identical to those of its congeneric living relative (*B. bison* and *H. sapiens* respectively), and, secondly, distinctly different from those of related genera (*Bos* in the case of *Bison*, and *Pan*, *Gorilla* and *Pongo* in the case of *Homo*). The sequences cannot, it seems, allow species to be differentiated, but they can resolve generic membership.

Let's make a big assumption and imagine that osteocalcin sequences can be recovered from (say) Mesozoic dinosaur taxa. Unfortunately, we're not going to be able to resolve synonymy at the specific level, nor discover cryptic species in the same way that DNA analysis can. But we would be able to resolve matters of generic membership: whether two genera are really synonymous for example (e.g., *Tyrannosaurus* and *Tarbosaurus*), and whether a *nomen dubium* represents something distinct, or is just a non-diagnostic remnant of another taxon. Such knowledge would have a huge impact on diversity studies. I think this has the potential to be something pretty big, IF osteocalcin really does prove as resistant, accessible and widespread among fossils as has been implied.

Where do we go from here, and are we on the brink of a revolution in palaeodiversity studies?

Refs - -

- Alroy, J. 2002. How many named species are valid? *Proceedings of the National Academy of Sciences* 99, 3706-3711.
- Alström, P., Mild, K. & Zetterström, B. 2003. *Pipits and Wagtails of Europe, Asia and North America.* Christopher Helm (London).
- Benton, M. J. 2005. The discovery pattern of dinosaurs … and how many more species are to be found? In Barrett, P. M. (ed) *53rd Symposium of Vertebrate Palaeontology and Comparative Antomy, Abstracts.* The Natural History Museum (London), p. 8.
- Collins, J. T. 1991. Viewpoint: a new taxonomic arrangement for some North American amphibians and reptiles. *Herpetological Review* 22 (2), 42-43.
- Glaw, F. Vences, M. 2002a. A new cryptic frog species of the *Mantidactylus boulengeri* group with a divergent vocal sac structure. *Amphibia-Reptilia* 23, 293-304.
- Glaw, F. & Vences, M. 2002b. A new sibling species of the anuran subgenus *Blommersia* from Madagascar (Amphibia: Mantellidae: *Mantidactylus*) and its molecular phylogenetic relationships. *Herpetological Journal* 12, 11-20.

- Glaw, F., Vences, M., Andreone, F. & Vallan, D. 2001. Revision of the *Boophis majori* group (Amphibia: Mantellidae) from Madagascar, with descriptions of five new species. *Zoological Journal of the Linnean Society* 133, 495-529.
- Mayer, F. & von Helversen, O. 2001. Cryptic diversity in European bats. *Proceedings of the Royal Society of London B* 268, 1825-1832.
- Molnar, R. E. 1990. Variation in theory and in theropods. In Carpenter, K. & Currie, P. J. (eds) *Dinosaur Systematics: Approaches and Perspectives* (Cambridge University Press, Cambridge), pp. 71-79.
- Nielsen-Marsh, C. M., Ostrom, P. H., Gandhi, H., Shapiro, B., Cooper, A., Hauschka, P. V. & Collins, M. J. 2002. Sequence preservation of osteocalcin protein and mitochondrial DNA in bison bones older than 55 ka. *Geology* 30, 1099-1102.
- Nielsen-Marsh, C. M., Richards, M. P., Hauschka, P. V., Thomas-Oates, J. E., Trinkaus, E., Pettitt, P. B., Karavanic, I., Poinar, H. & Collins, M. J. 2005. Osteocalcin protein sequences of Neanderthals and modern primates. *Proceedings of the National Academy of Sciences* 102, 4409-4413.
- Olsson, U., Alström, P., Ericson, P. G. P. & Sundberg, P. 2005. Non-monophyletic taxa and cryptic species – evidence from a molecular phylogeny of leaf-warblers (*Phylloscopus*, Aves). *Molecular Phylogenetics and Evolution* 36, 261-276.
- Parra-Olea, G., Garcia-Paris, M., Papenfuss, T. J. & Wake, D. B. 2005. Systematics of the *Pseudoeurycea bellii* (Caudata: Plethodontidae) species complex. *Herpetologica* 61, 145-158.
- Peters, J. L. 1951. *Check-List of birds of the world. Volume VII.* Museum of Comparative Zoology, Cambridge, Massachusetts.
- Robertson, C. J. R. & Nunn, G. B. 1998. Towards a new taxonomy for albatrosses. In Robertson, G. & Gales, R. (eds) *Albatross Biology and Conservation.* Surrey Beatty (Sydney), pp. 13-19.
- Sibley, C. G. & Monroe, B. L. 1991. *Distribution and Taxonomy of Birds of the World.* Yale University Press, New Haven.
- Vallan, D., Vences, M. & Glaw, F. 2003. Two new species of the *Boophis mandraka* complex (Anura, Mantellidae) from the Andasibe region in eastern Madagascar. *Amphibia-Reptilia* 24, 305-319.
- Wüster, W. & Thorpe, R. S. 1994. *Naja siamensis*, a cryptic species of venomous snake revealed by mtDNA sequencing. *Experientia* 50, 75-79.
- Zink, R. M. 1996. Bird species diversity. *Nature* 381, 566.
- Zink, R. M. 2004. The role of subspecies in obscuring avian biological diversity and misleading conservation policy. *Proceedings of the Royal Society of London B* 271, 561-564.

Chapter 15: When salamanders invaded the Dinaric Karst: convergence, history, and reinvention of the troglobitic olm

One of the most unusual and interesting of amphibians has to be the Olm (*Proteus anguinus*), an unusual long-bodied cave-dwelling salamander from SE Europe. Olms were the first specialised cave-dwelling animals (so-called stygobionts or troglobites) to be discovered, they were traditionally identified as dragon larvae by local people, and they remain mysterious and the source of controversy, debate and discovery. I've had a special affinity for olms since seeing them (live) in the former Yugoslavia in 1987, and after a colleague published a brief article on them in 2004 I ended up compiling and publishing my olm-related thoughts. In the interests of re-cycling that text I reproduce it here, in updated form.

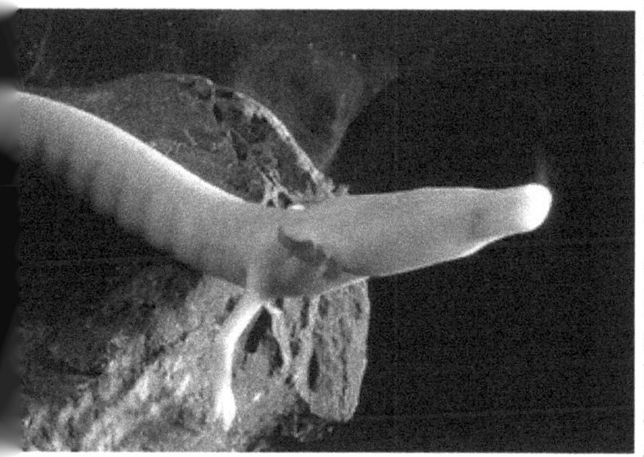

Fig. 49. White olm. Photo by Arne Hodalic.

What might be the most fascinating fact concerning olms is the most poorly-known and least mentioned one: the 1986 discovery of a surface-dwelling olm, described in 1994 by Boris Sket and Jan Willem Arntzen. So olms aren't just 'unusual long-bodied *cave-dwelling* salamanders' – they now exist in two forms, the cave-dwelling White olm *Proteus anguinus anguinus* and the surface-dwelling Black olm or Brown olm *P. a. parkelj*. Unlike the unpigmented nominal form with its skin-covered eyes, *P. a. parkelj* (presently known only from Bela Krajina in SE Slovenia) is dark brown or black and has externally visible (albeit small) eyes (Fig. 50). Because White olms produce melanin when kept in sunlight (and are thus not albinistic as sometimes implied), the difference in colour between the two forms is not unexpected. However, there are also other, more important differences separating the two. *P. a. parkelj* differs from the nominal form in also having a proportionally shorter, broader and more muscular head, fewer teeth, a proportionally longer body and a proportionally shorter tail and limbs (Sket & Arntzen 1994).

Fig. 50. Black olm. Photo by Arne Hodalic.

Most of the features which distinguish *P. a. parkelj* from *P. a. anguinus* are plesiomorphies [= features not unique to olms, but present also in related salamanders] and hence *P. a. parkelj* may be the ancestor of the White olm. Having said that, one of the most interesting contentions made recently about olms (Sket 1997) is that the different cave-dwelling olm populations may have evolved independently from different ancestral populations. If this is correct it may be that the different White olm populations represent different species which resemble one another by convergent evolution, and which have partly or mostly fused as they have met up within the Dinaric karst system. Sket (1997) thought that morphological and genetic differences observed among olms might provide support for this view and, ironically, if correct it would mean that several old species names proposed for different cave-dwelling olms might need to be resurrected. Fitzinger (1850) named seven new olm species within his genus *Hypochthon* (*H. zoissii, H. schreibersii, H. freyeri, H. haidingeri, H. laurentii, H. xanthostictus* and *H. carrarae*), though given that the type localities for some of them were just a few kilometres apart, it's unlikely that they really were distinct taxa.

It's worth saying that olms almost certainly aren't ancient relicts, or living fossils. In fact, they must be young and recently evolved. Why? During the Pleistocene, the Dinaric area was so close to areas that were fully glaciated that temperature there must have been at or below freezing. This is far too cold for olms, which require temperatures of 6-18° C for their eggs and larvae to develop (and toward the upper end of that range is best). Furthermore, karstification and the development of underground streams only began in the Dinaric region during the late Pliocene at the earliest, apparently. In view of these problems, olms either (1) survived in surface waters in the region, where summer temperatures were just about tolerable (but where winter temperatures would have made life difficult), or (2) moved into the area from a warmer, southerly refuge (Griffiths 1996, Sket 1997). It isn't yet known which was the case: more research is needed. Whatever, troglobitic olm populations must have evolved within the last 10,000 years or so, and presumably the specialised troglobitic morphology of living olms evolved during this time. Similarly recent invasions of cave systems appear to have occurred among various troglobitic fishes.

As mentioned earlier, olms were the earliest troglobites to be discovered. While it's been stated on occasion that the species was discovered as recently as 1875 (Laňka & Vít 1985, Keeling 2004), olms first became widely known in 1689 when Baron Johann Weichard Valvasor wrote about the animals in his book on the Yugoslavian province of Carniola (on which see below). However, it wasn't until the mid-1700s that the animals become the subject of proper scientific debate. At this time Slovenian scientist Giovanni Scopoli 'discovered' olms and realized just how extraordinary they were (Scopoli 1772). He planned to describe the animal scientifically and enhance his reputation by doing so.

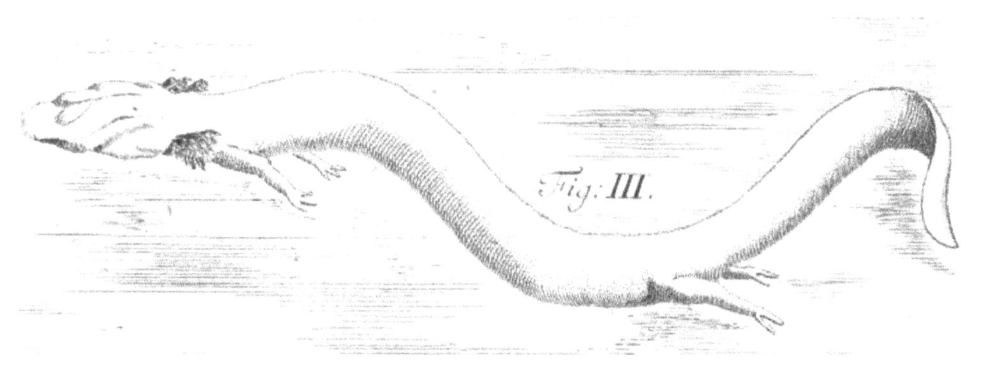

Fig. 51. Laurenti's olm illustration from 1768.

We know that Scopoli sent pictures of olms to Carl von Linné and that Linné and Scopoli disagreed as to whether the animals were a distinct new genus (Scopoli's view), or the juveniles of something else (Linné's view). However, the Austrian anatomist J. N. Laurenti became very interested in olms at the same time as Scopoli (apparently because of a specimen Scopoli had sent to one of the Laurenti's friends) and, in 1768, published the first scientific description of the species (Laurenti 1768). This is the 'official' date of the White olm's scientific discovery. Laurenti's choice of generic name for the olm (*Proteus*) is based on the Greek god Proteus but it is not Proteus' shape-shifting ability that Laurenti had in mind, but rather his status as shepherd of sea creatures. Laurenti's work on the olm did not actually become that well known and it was Karl von Schreibers's work of the 1790s and early 1800s which made olms well known among scientists.

Olm distribution has been the subject of much confusion and speculation. Presently, olms are restricted to the Dinaric Karst, a region that extends from the Soca (formerly the Isonzo) River (near Trieste) in SE Friuli-Venezia Giulia, Italy to the Trebišnica River in eastern Herzogovina. In between Italy and Herzogovina, olms also occur in southern Slovenia, southern Croatia and parts of Bosnia. Little known is that the species has been recorded from localities in France (Moulis) and Germany (Harz). These extralimital records are all apparently due to human introduction however. They are also found in the Parolini Grotto, Vicenza, northern Italy, but their presence here is due to human introduction also. Of further interest, the locality mentioned by Valvasor (1689) – the spring of Lintvern, near Vrhnika – is actually outside of the Dinaric Karst, and is unlike the other areas inhabited by olms in geology and geomorphology. It seems that Valvasor made the logical (but incorrect) assumption that Lintvern (which is a garbled form of the German word Lindwurm, meaning dragon) was so named because it was the source of olms (which were fancifully regarded as dragon larvae at the time).

Keeling (2004) implied that Carniola (note: not Carinola) might be the Italian part of the olm's range and also wondered if Carniola might still be part of Austria. Carniola is today called Kranjska and was controlled by Austrian royalty until 1918 (consequently, the ruling classes there spoke German until the 20th century). It is today part of central Slovenia and is thus not either the Italian part of the olm's range, nor an Austrian extension of the species' range.

Fig. 52. White olm. Photo by Arne Hodalic.

Bizarrely, olms were traded during Victorian times as exotic pets and were apparently available in Britain as such (which raises the remote possibility that they might have been introduced to British cave systems in the same way that they were in French, German and Italian ones). During the 1950s it was reported that olms were present in the Carpathian karst of eastern Serbia, and in 1960 a team of speleologists from Ljubljana led an expedition to the region to investigate this possibility. They didn't find any olms there, and nor has anyone else since.

Olms have been horrendously over-collected for scientific use and were also apparently collected by farmers for use as pig food. One of the greatest problems facing olms today is metal poisoning caused by industrial pollution and a number of populations have declined as a result of such. *P. a. parkelj* is under strict legal protection. Olms have been protected in Slovenia since 1949 at least: elsewhere in their range they are widely recognized as being deserving of protection.

Finally, regarding diet and breeding, olms apparently mostly detect their prey using chemical clues and the detection of water currents but they also possess electroreceptive organs in the head and thus presumably employ electroreception. Despite their vestigial nature, the eyes of White olms are not completely useless and are able to detect light. Olms appear to mostly prey on aquatic crustaceans but also eat snails and insect larvae. Captive specimens have eaten worms and adults may be cannibalistic on occasion. When I visited Postojina we were told that the olms on display were not fed both because their food proved hard to procure, and because they were quite able to survive for years without feeding. Indeed there was apparently a specimen kept at the Faculty of Biotechnology in Ljubljana which survived for an astonishing 12 years without food. Olms are long-lived, reaching sexual maturity between 7 and 14 years, and almost certainly ordinarily live for more than 50 years, though ages twice this have been suggested by some writers.

Refs - -

- Fitzinger, L. 1850. Ueber den *Proteus anguinus* der Autoren. *Sitz.-Ber. Akad. Wiss., Math.-naturw. Cl.* 5, 291-303.
- Griffiths, R. A. 1996. *Newts and Salamanders of Europe*. T & A D Poyser (London).
- Keeling, C. 2004. Olm. *Mainly About Animals* July 2004, 20-21.
- Laňka, V. & Vít, Z. 1985. *Amphibians and Reptiles*. Hamlyn (London).
- Laurenti, J. L. 1768. *Specimen Medicum Exhibens Synopsis Reptilium Emendatum*. Joan. Thomae (Vienna).
- Scopoli, J. A. 1772. *Annus Quintus Historico-Naturalis*. C. G. Hilscher (Lipsiae).
- Sket, B. 1997. Distribution of *Proteus* (Amphibia: Urodela: Proteidae) and its possible explanation. *Journal of Biogeography* 24, 263-280.
- Sket, B. & Arntzen, J. W. 1994. A black, non-troglomorphic amphibian from the karst of Slovenia: *Proteus anguinus parkelj* n. ssp. (Urodela: Proteidae). *Bijdr. Dierk.* 64, 33-53.
- Valvasor, J. W. 1689. *Die Ehre des Herzogthums Crain*. W. M. Endtner, Nuernberg.

Chapter 16:
Pleistocene refugia and late speciation: are extant bird species older than we mostly think?

Climatic changes affect the distributions of organisms. This assertion is self-evident, not controversial, and indeed observable within a human lifetime. So given that the planet has experienced major fluctuations in climate within the recent geological past, it follows that well-vegetated habitats were fragmented during the dry cycles of the Pleistocene, and that previously contiguous animal populations became divided. For the purposes of this discussion, this fragmentation had two results: (1) that populations became restricted to refugia – that is, islands of surviving forested habitat; and (2) that speciation was encouraged and accelerated during this time (driving so-called Late Pleistocene Origins, and resulting in the LPO model).

So that's the 'glacial refugium' hypothesis (Rand 1948, Stewart & Lister 2001).

I feel that this view of Pleistocene environ-

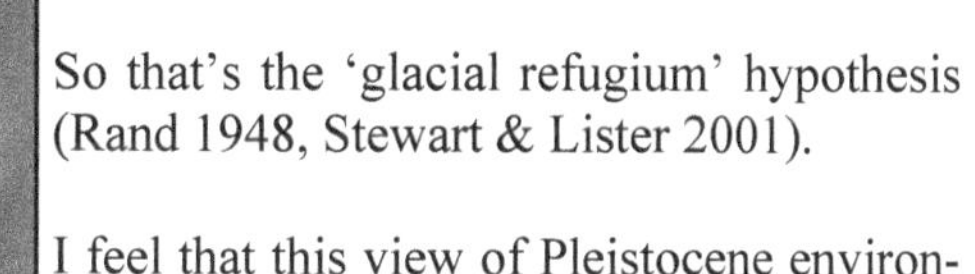

Fig. 53. Modern birds, like this House sparrow *Passer domesticus*, are usually thought to be young species less than a couple of million years old. But is this wrong?

mental change is logical and, at face value, well supported. It's become well accepted to the extent that it's become the stuff we find in textbooks (e.g., Wiens 1991). But interesting things are happening, and it's after reading recent papers on rattlesnake phylogeography that I decided to write about this. First, I went to the bird literature, as there is an awful lot of it on LPO. If you're most interested in the rattlesnakes, prepare to be disappointed.

As a nerdy teenager reading all I could on animals, the view I grew up with was that extant birds had speciated during the Pleistocene, and that passerines in particular provided excellent support for LPO. Indeed, the LPO hypothesis has been mostly driven by studies on passerine speciation (Brodkorb 1960). However, this view has come under attack. Genetic data on speciation rates in passerines, published within the last ten or so years, does NOT support the LPO model.

By looking at recently diverged taxa among North American passerine clades (including grackles, tits, parulids, thrashers and icterids), Klicka & Zink (1997) found that genetic divergence rates actually indicated splits rather older than they'd need to be to satisfy the LPO model: on average, the data suggested Late Pliocene divergence times occurring at about 2.5 million years ago. They determined this by assuming molecular clock rates of 2% per million years however, and herein lies the Achilles heel of their study, as there is considerable doubt as to the idea that genetic changes continue at a clock-like rate. They acknowledged this, and only compared taxa thought to have similar mutation rates. Arbogast & Slowinski (1998) produced a rebuttal arguing that the 2% divergence rate was inaccurate and that Klicka & Zink's speciation dates were therefore erroneous, but reanalysis still didn't produce inferred speciation rates young enough to satisfy the LPO model (Klicka & Zink 1998). As Klicka & Zink (1997) state "These results contradict the expectations of the LPO model. Overall, these data reflect a protracted history of speciation throughout the Pleistocene and Pliocene" (pp. 1666-1667), and they termed the LPO model 'a failed paradigm'. Other genetic studies have also found that extant bird species are actually pretty old (e.g., Zink & Slowinski 1995, Avise & Walker 1998, Avise *et al.* 1998).

This is very interesting from the palaeontological perspective: it means that by far the majority of passerines aren't modern/Pleistocene novelties, but have actually been around for a while. Indeed the only taxon found by Klicka & Zink (1997) to be a novelty of this sort is the Timberline sparrow *Spizella taverneri*: it seems to have diverged from *S. breweri* about 35,000 years ago (incidentally, American 'sparrows' aren't sparrows at all, but buntings).

What do the fossils say about this? Well that's an interesting question, as different palaeornithologists have held conflicting perspectives on this issue. Taken at face value, the fossil record seems mostly to support the LPO model, given that Pleistocene bird fossils are unique to the Pleistocene. Brodkorb (1960) concluded that this was for real, and that few bird species crossed the end-Pleistocene boundary.

Conversely though, others have argued that Pleistocene 'species' aren't demonstrably distinct from Holocene ones – they might only be given different species names because of convention. A review of Pleistocene 'species' from Europe showed that the characters used to differentiate Pleistocene 'species' from modern ones were mostly vague and unsatisfactory assertions about size or robustness (Stewart 2002). One of the most important factors resulting in the recognition of a new Pleistocene taxon proved to be its age... yes, that's right: because it was from the Pleistocene, it must have been a new species. However, this is perhaps an unduly negative view. Tyrberg (2002) found that "few, if any, avian species are of Late Pleistocene age [DN: meaning that they're older] while at least half of the extant Palearctic bird species have their origins in the Pliocene" (p. 281). If he's right, then the fossil data is in agreement with what the molecular workers report, and we should not be surprised when extant species are reported from the Pliocene, as they sometimes are.

But – hold on – extant species haven't just been reported from the Pliocene, but also from the Miocene. Should we be taking this seriously too? Tyrberg (2002) reminds us of the case of *Archaeotrogon venustus*, an archaeotrogonid that persists in the fossil record for something like 14 million years. Assuming that all the fossils identified as belonging to this species really do represent the same animal, this is the longest range recorded for a bird. It indicates that "reports of extant species in the Late Miocene should not be rejected out of hand as is usually done" (Tyrberg 2002, p. 287).

Now, this makes things even more interesting. I'm thinking New Guinea. Why? Many bird genera there – particularly paradisaeids and ptilonorhynchids – have bizarre disjunct distributions. The three similar *Paradisaea* species *P. rubra*, *P. guilielmi* and *P. decora* are found in the Moluccas (to the west of New Guinea) and in extreme eastern New Guinea, but nowhere in between. The similar astrapia species *Astrapia nigra* (Fig. 54) and *A. rothschildi* live in the Arfak and Tamrau mountains (to the extreme west) and on the Huon Peninsula (to the extreme east), respectively. How can these disjunct distributions be explained? Heads (2001a) argued that the birds must be the products of vicariance: the areas where they occur were formerly close, but as the microterranes moved, the birds have simply been ultra-sedentary and gone with them.

Fig. 54. Arfak astrapia *Astrapia nigra*. Illustration by Richard Bowdler Sharpe.

While there are very good reasons for thinking that these birds really are this sedentary (while many birds are great at dispersing, a great many others simply aren't [Diamond 1981]), I used to think that this just couldn't be right as there was no way the bird genera, let alone the species, could possibly be old enough. Well, now I'm not so sure. The key tectonic events seem to have occurred in the Miocene, and Heads concluded that the original non-disjunct distribution of the birds must really have dated from this time. Passerines aren't the only New Guinean taxa with these distributions by the way – it's present in plants and other groups too (Heads 2001a, b, c, d, 2002).

My plan originally was to discuss how Neotropical rattlesnake phylogeography has actually supported the concept of glacial refugia, then to go from there to the proposed Pleistocene fragmentation of Amazonia. But because birds don't support the LPO model, they have ended up providing no support for this model, and in fact work on the timing of avian speciation has gone hand-in-hand with criticisms of the refugium theory.

Refs - -

- Arbogast, B. S. & Slowinski, J. B. 1998. Pleistocene speciation and the mitochondrial DNA clock. *Science* 282, 1955a.
- Avise, J. C. & Walker, D. 1998. Pleistocene phylogeographic effects on avian populations and the speciation process. *Proceedings of the Royal Society of London B* 265, 457-463.
- Avise, J. C., Walker, D. & Johns, G. C. 1998. Speciation durations and Pleistocene effects on vertebrate phylogeography. *Proceedings of the Royal Society of London B* 265, 1707-1712.
- Brodkorb, P. 1960. How many bird species have existed? *Bulletin of the Florida State Museum, Biological Sciences* 5 (3), 41-56.
- Diamond, J. 1981. Flightlessness and fear of flying in island species. *Nature* 293, 507-508.
- Heads, M. 2001a. Birds of paradise, biogeography and ecology in New Guinea: a review. *Journal of Biogeography* 28, 893-925.
- Heads, M. 2001b. Birds of paradise (Paradisaeidae) and bowerbirds (Ptilonorhynchidae): regional levels of biodiversity and terrane tectonics in New Guinea. *Journal of Zoology* 255, 331-339.
- Heads, M. 2001c. Regional patterns of biodiversity in New Guinea plants. *Botanical Journal of the Linnean Society* 136, 67-73.
- Heads, M. 2001d. Birds of paradise, vicariance biogeography and terrane tectonics in New Guinea. *Journal of Biogeography* 29, 261-283.
- Heads, M. 2002. Regional patterns of biodiversity in New Guinea animals. *Journal of Biogeography* 29, 285-294.
- Klicka, J. & Zink, R. M. 1997. The importance of recent ice ages in speciation: a failed paradigm. *Science* 277, 1666-1669.
- Klicka, J. & Zink, R. M. 1998. Pleistocene speciation and the mitochondrial DNA clock: response to Arbogast & Slowinski. *Science* 282, 1955a.
- Rand, A. L. 1948. Glaciation, an isolating factor in speciation. *Evolution* 2, 314-321.
- Stewart, J. R. 2002. The evidence for the timing of speciation of modern continental birds and the taxonomic ambiguity of the Quaternary fossil record. In Zhou, Z. & Zhang, F. (eds). *Proceedings of the 5th Symposium of the Society of Avian Paleontology and Evolution.* Science Press (Beijing), pp. 259-280.

- Stewart, J. R. & Lister, A. M. 2001. Cryptic northern refugia and the origins of the modern biota. *Trends in Ecology & Evolution* 16, 608-613.
- Tyrberg, T. 2002. Avian species turnover and species longevity in the Pleistocene of the Palearctic. In Zhou, Z. & Zhang, F. (eds). *Proceedings of the 5th Symposium of the Society of Avian Paleontology and Evolution.* Science Press (Beijing), pp. 281-289.
- Wiens, J. A. 1991. Evolutionary biogeography. In Brooke, M. & Birkhead, T. (eds) *The Cambridge Encyclopedia of Ornithology.* Cambridge University Press (Cambridge), pp. 156-161.
- Zink, R. M. & Slowinski, J. B. 1995. Evidence from molecular systematics for decreased avian diversification in the Pleistocene Epoch. *Proceedings of the National Academy of Sciences* 92, 5832-5835.

Chapter 17:
The late survival of *Homotherium* confirmed, and the Piltdown cats

Some of the most fascinating, illuminating insights into extinct animals come from data recorded by the people who saw them. We might be sad that the Pleistocene megafauna are gone, but at least our ancestors painted, carved and sculpted representations of them. In 1896, in the French cave of Isturitz, a 16-cm long statuette of a big cat was discovered. Initially interpreted as a representation of a cave lion, it was reinterpreted by Vratislav Mazak (1970) as more likely being a depiction of the sabre-tooth *Homotherium latidens* (a species sometimes dubbed the scimitar cat). Like a homothere, but unlike a lion, the statuette (which has since been lost) has a short tail and a deepened lower jaw. If the statuette is meant to depict *Homotherium*, it provides us with some new information on the life appearance of this cat, as it appears to be decorated with small spots, and to have a pale underside (Fig. 55). As Rousseau (1971a, 1971b) described, there are also other Palaeolithic pieces of cave art that appear to depict homotheres.

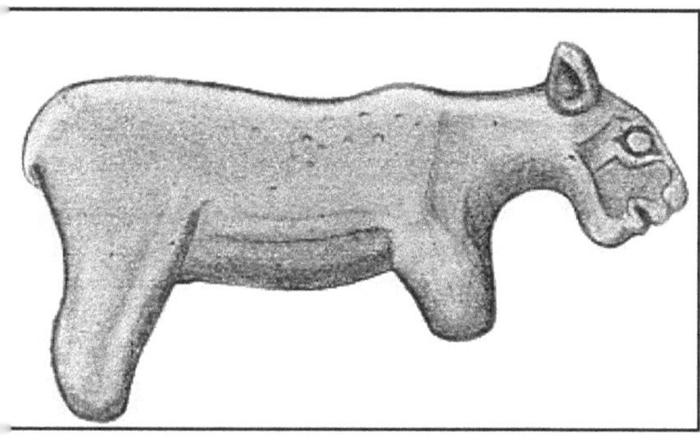

Fig. 55. The Isturitz statuette.

The problem though is that the Isturitz statuette (and other pieces of evidence) is somewhere around 30,000 years old, and the last accepted datum for skeletal material of *Homotherium* in Europe is 300,000 years BP (Adam 1961) [though see below]. This significant discrepancy therefore suggests that *Homotherium* survived in Europe for much later than thought but, given that this has until now been based on artwork, the area has remained controversial. As Shuker (1989) wrote: "Were the above works of Palaeolithic art nothing more than inaccurate or idealized depictions of cave lions, or do they comprise genuine proof that the extraordinary scimitar cat was a contemporary of our ancestors for a far longer period of time than hitherto believed?".

A new young homothere record

In March 2000 the fishing vessel UK33 trawled a partial felid lower jaw from an area SE of the Brown Bank in the North Sea, an area previously known for yielding Pleistocene and Holocene fossil mammals. As described by Jelle Reumer *et al.* (2003), the jaw is from a *Homotherium latidens*, and what is especially significant is that radiocarbon analysis dates it to 28,000 years BP. As Reumer *et al.* note, this is about the same age as the Isturitz statuette and therefore confirms the long-suspected late survival of this felid in Europe.

Incidentally, the climate in northern Europe at this time would have been quite harsh - the Devensian Glaciation was at its height between 25,000 and 15,000 years BP, and at this time northern Britain as far south as Yorkshire was covered by an ice sheet. Cold tundra and steppe environments occurred to the south and east of this ice sheet, and only cold-tolerant species could have lived in the area now occupied by the North Sea. Reindeer were living in Cambridgeshire, Polar bears in London, and Musk ox in Wiltshire. *Homotherium latidens* must also have therefore been a cold-tolerant species. Given that *Homotherium* species also dwelt in temperate and tropical environments (in Asia homotheres are known as far south as Java), this was clearly a highly adaptable felid.

The Piltdown cats

Prior to Reumer *et al.*'s discovery there were a number of British homothere fossils which were initially regarded as coming from late glacial deposits, and thus being somewhere around 13,000-11,000 years BP in age (i.e., as young as the youngest possible age for the youngest American material). Most famously they include a single canine from Robin Hood Cave, the largest cave of the Creswell Crag complex at Derbyshire, discovered in 1876. Describing the tooth in 1877, William Boyd Dawkins, the pioneering geologist and expert on Palaeolithic man, suggested that it may have been introduced into the cave by humans, as it appeared to bear both the marks of a flint tool, and an incomplete perforation at its base. On balance though, Dawkins concluded that the tooth suggested late survival of *Homotherium* in Britain. This idea has been mentioned by other workers and it led Kurtén (1968) to suggest that *H. latidens* survived in Britain for far longer than it did in mainland Europe, or in other words that Britain acted as a refugium for this disappearing species.

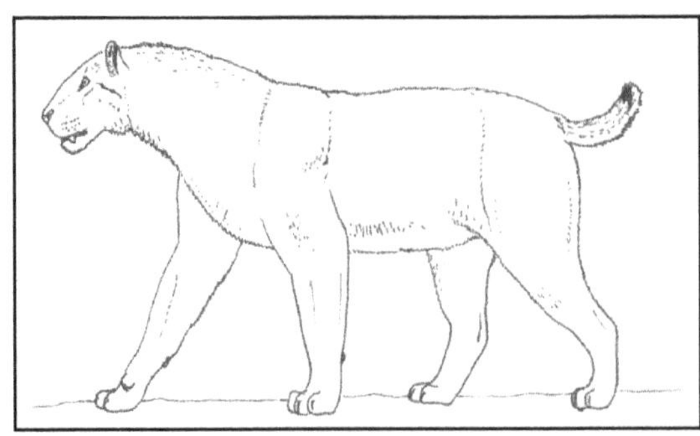

Fig. 56. Life reconstruction of *Homotherium*, based on an earlier illustration by Mauricio Anton.

Although it may have been separated from mainland Europe during one or more of the Pleistocene interglacials (namely during part of the Ipswichian Interglacial, between c. 130,000-70,000 years BP), the English Channel did not flood until about 9000 years before present, so any homothere living in Britain between 13,000 and 11,000 years ago could still have walked to mainland Europe (Stuart 1974, Yalden 1982).

It is of further interest to note that, had a hypothetical homothere population become isolated in interglacial Britain, it may only have lasted for about 1000 years before become extinct due to inbreeding. Recent modelling work on population viability in large Pleistocene carnivorans (O'Regan *et al.* 2002) has shown that even glacial refugia the size of the Iberian and Italian peninsulas were not big enough for large felids to survive in when these populations became isolated, as they apparently did during the Pleistocene glaciations. Lack of space during these times may therefore have promoted extinction, an issue that is particularly poignant today as large carnivorans find themselves restricted to increasingly smaller islands of suitable habitat.

Returning to the Robin Hood Cave tooth, recent work indicates that Dawkins's initial suspicions were right. The fact that the tooth had been altered by humans indicates that it probably was traded and carried around by them long after its original owner had died (Charles & Jacobi 1994). Furthemore, the specimen was apparently discovered on one of the four days on which the senior archaeologist in charge of the site – Tom Heath – was absent. Consequently it is not surprising that hoaxing has been suggested at various times, and Yalden (1999) compared the Robin Hood Cave homothere to the Piltdown fossils. This remains an unproven assertion however, and Kenneth Oakley's (1980) discovery that the tooth differs in its fluorine, uranium and nitrogen content from all other British homothere fossils has been used as evidence both for and against its being a hoax. Given these problems, however, it has been recommended that this record be ignored.

British homotheres are also known from the early Pleistocene site of Dove Holes near Buxton, Derbyshire, a site that also yielded giant hyaenas, straight-tusked elephants and southern mammoths (Dawkins 1903), but is today occupied by a municipal rubbish dump (Yalden 1999). Middle Pleistocene British homotheres are known from the cavern infill site near Westbury-sub-Mendip, Somerset (Bishop 1982). Finally, the Kent's Cavern teeth - initially thought to be late Pleistocene - come from a cave that also contains older Pleistocene fossils, and it is now thought that they are also middle Pleistocene.

Refs - -

- Adam, K. D. 1961. Die Bedeutung der pleistozanen Saugetier-Faunen Mitteleuropas fur die Geschichte des Eiszeitalters. *Stuttgarter Beitrage zur Naturkunde* 78, 1-34.
- Bishop, M. J. 1982. The mammal fauna of the early Middle Pleistocene cavern infill site of Westbury-sub-Mendip Somerset. *Special Papers in Palaeontology* 28, 1-108.
- Charles, R. & Jacobi, R. M. 1994. The Lateglacial fauna from Robin Hood Cave, Cresswell: a re-assessment. *Oxford Journal of Archaeology* 13, 1-32.
- Dawkins, W. B. 1903. On the discovery of an ossiferous cavern of Pliocene age at

Dove Holes, Buxton (Derbyshire). *Quarterly Journal of the Geological Society, London* 59, 105-133.

- Kurtén, B. 1968. *Pleistocene Mammals of Europe*. Weidenfeld & Nicolson (London).
- Mazak, V. 1970. On a supposed prehistoric representation of the Pleistocene scimitar cat, *Homotherium* Farbrini, 1890 (Mammalia; Machairodontinae). *Zeitschrift fur Saugertierkunde* 35, 359-362.
- Oakley, K. 1980. Relative dating of the fossil hominids of Europe. *Bulletin of the British Museum (Natural History), Geology* 34, 1-63.
- O'Regan, H. J., Turner, A. & Wilkinson, D. M. 2002. European Quaternary refugia: a factor in large carnivore extinction? *Journal of Quaternary Science* 17, 789-795.
- Reumer, J. W. F., Rook, L., Van Der Borg, K., Post, K., Mol, D. & De Vos, J. 2003. Late Pleistocene survival of the saber-toothed cat *Homotherium* in northwestern Europe. *Journal of Vertebrate Paleontology* 23, 260-262.
- Rousseau, M. 1971a. Un félin à canine-poignard dans l'art paléolithique? *Archéologia* 40, 81-82.
- Rousseau, M. 1971b. Un machairodonte dans l'art aurignacien? *Mammalia* 35, 648-657.
- Shuker, K. P. N. 1989. *Mystery Cats of the World*. Robert Hale (London).
- Stuart, A. J. 1974. Pleistocene history of the British vertebrate fauna. *Biological Reviews* 49, 225-266.
- Yalden, D. W. 1982. When did the mammal fauna of the British Isles arrive? *Mammal Review* 12, 1-57.
- Yalden, D. W. 1999. *The History of British Mammals*. Poyster Natural History, London.

Chapter 18:

It's not a rhinogradentian: it's the most fantastic jerboa, *Euchoreutes*

It's funny how things work out. Today I am obsessed with rodents. Why? Most of my day was spent clearing out an old loft, and while rummaging through decades of accumulated rubbish I came across multiple copies of old Brooke Bond picture card albums, and among them one of my favourites: Tunnicliffe's *Asian Wild Life* (Fig. 57). Brooke Bond pictures cards were given away free inside boxes of tea (the tea-producing branch of the company later became known as PG Tips) and, for a small fee, collectors could send off for an album. Hugely influential to young people that grew up in tea-drinking households during the 1960s and 70s, many of the series were devoted to natural history, and they are fondly remembered by many people who work today in the biological sciences. They explain my fascination with the artwork of Peter Scott, Charles Tunnicliffe and Maurice Wilson.

Fig. 57. The front cover of the Brooke Bond picture card album *Asian Wild Life*, illustrated by Charles F. Tunnicliffe.

While, mostly, I looked after the albums that I inherited from my mother – who collected the cards herself as a girl – there were a few that I unfortunately defaced and mutilated, *Asian Wild Life* among them. So today I'm happy to have back in my hands not one, but two, pristine, completed albums. Like several of the Brooke

Bond picture cards series, *Asian Wild Life* was both written and illustrated by the fantastic Charles F. Tunnicliffe (1901-1979). And there is always one picture in particular that fascinated me, and today still does: it's Tunnicliffe's painting (reproduced below) of two Yarkand jerboas *Euchoreutes naso*, bounding together across the steppes of north-west China (Fig. 58).

Euchoreutes has to be one of the oddest-looking rodents and, years later, when I learnt about the rhinogradentians, I wondered if *Euchoreutes* wasn't really a jerboa at all, but in fact a wayward rhinogradentian, perhaps related to the Earwing *Otopteryx volitans*. Even the binomial – *Euchoreutes naso* – is suggestive of some link with rhinogradentians given that the latter group includes the nasobemes (genus *Nasobema*). Like earwings, *Euchoreutes* has ridiculously enormous ears, and its alternative name is the Long-eared jerboa. If anything, Tunnicliffe's painting actually doesn't make the ears appear large enough: in photos, the ears look to be about as long as the entire body.

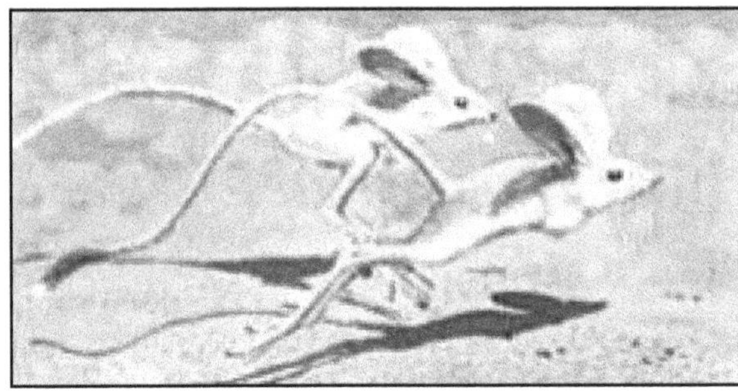

Fig. 58. Yarkand jerboas as illustrated by Charles F. Tunnicliffe.

And in body length, *Euchoreutes* is just 70-90 mm long. That's small, but not as small as the Baluchistan pygmy jerboa *Salpingotus michaelis*: with a body length of 36-47 mm it's the smallest living rodent (it's also relatively new to science, having only been discovered in 1966 and named in 1973). At the other end of the scale, some species of *Allactaga* (the four- and five-toed jerboas) exceed 260 mm in body length, and can then have a tail over 300 mm long. Getting back to *Euchoreutes*, it's odd not just for its large ears, but also for its unusually long snout. It does however resemble most other jerboas in having proportionally small forelimbs and tremendously elongate hindlimbs.

Like all jerboas (well… nearly all: read on), *Euchoreutes* has fused metatarsals. Is metatarsal fusion a synapomorphy for the group? There's a problem with that: the Five-toed dwarf jerboa *Cardiocranius paradoxus* lacks metatarsal fusion. Is this because it's the most basal jerboa, because it exhibits a character reversal, or because it's not a jerboa at all? While few phylogenetic studies incorporate it (it is a very obscure and little-studied species), it is usually implied in classifications that it's down at the base of the jerboa clade (properly called Dipodidae).

Though pedal digits I and V are reduced in *Euchoreutes*, they are still present. This contrasts with the dipodine jerboas *Paradipus*, *Dipus*, *Stylodipus*, *Eremodipus* and *Jaculus*, all of which lack lateral digits and are tridactyl. Their elongate, fused metatarsi thus bear three distinct distal condyles and look, at least superficially, remarkably like the tarsometatarsi of birds. This similarity has not been lost on ornithologists (Rich 1973) and is a remarkable case of convergent evolution. If the proximal end of the metatarsus were broken off (and this bit is the give-away, as it of course shows the presence of tarsals charactestically mammalian in form and number), I suspect that even some experienced zoologists would be fooled into misidentifying a jerboa metatarsus as an avian one. Sadly I don't have many jerboa leg skeletons lying around so cannot test this idea. Incidentally, most of the cervical vertebrae in jerboas are fused together as well, and in some dwarf jerboas the first three dorsal vertebrae are also fused together, and to the fused cervicals. I don't know why this is, but it might be to prevent dislocation or jarring during the violent acceleration and deceleration incurred during leaping and bounding.

And on the subject of leaping and bounding, jerboa feet are clearly specialised for saltation (jumping). With body lengths of mostly around 100 mm, jerboas can cover about 3 m in a single leap. This is a neat and useful trick if you want to cross large distances on hot sand, but of course jerboas are mostly nocturnal, and the predominant function of saltation in jerboas is to move quickly away from predators. One species – the Rough-legged jerboa *Dipus sagitta* – exhibits particularly interesting predator-avoidance behaviour: it not only leaps from predators, but, as it leaps, grabs at over-hanging foliage with its teeth and forelimbs, and then clambers into the vegetation to hide (Hanney 1975).

Specialised as they are for impoverished steppes, sub-deserts and deserts, jerboas have apparently benefited from desertification in some regions (Duplaix & Simon 1977). This probably only applies to tolerant generalists among the group, however, and certainly doesn't work for *Euchoreutes*. It reportedly declined by about 50% during the 1990s (Nowak 1999) and is regarded as endangered.

Phylogenetic studies demonstrate that *Euchoreutes* really is a jerboa, and not a rhinogradentian, and it's traditionally been allocated its own 'subfamily' called Euchoreutinae Lyon, 1901 within the jerboa family Dipodidae Fischer de Waldheim, 1817. Whether *Euchoreutes* is actually a member of either of the two dipodid clades that have been recognised in phylogenetic studies of this group – Dipodinae and Allactaginae (Shenbrot 1992) – remains uncertain. One study of dipodid phylogeny based on cranial characters (Dempsey 1991) didn't include *Euchoreutes* as no skulls were available for examination, which isn't surprising given that only a handful of specimens are present in museums worldwide (Nowak 1999). Classifications have generally listed Euchoreutinae as separate from Dipodinae and Allactaginae, but only because

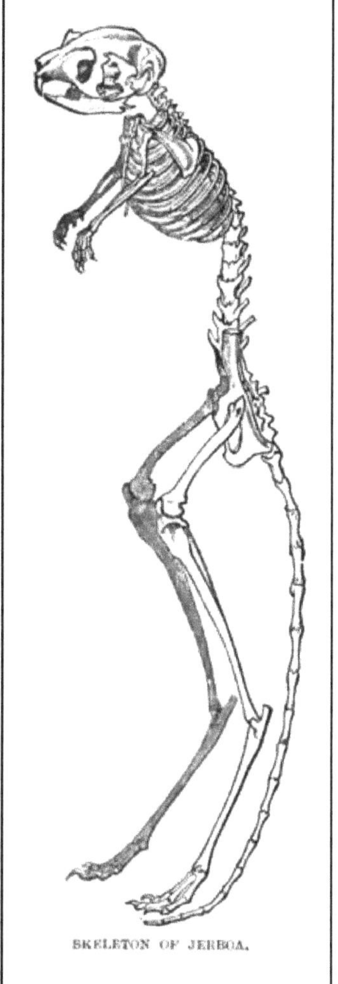

Fig. 59. Jerboa skeleton from Richard Lydekker's *The Royal Natural History, Volume 3*. Lydekker didn't identify the species shown here, but it's most likely one of the *Jaculus* species.

the 'subfamily' rankings demand that each be treated as equivalent in rank. So exactly how does *Euchoreutes* fit into dipodid phylogeny? That's a good question, and I'd be interested to know if it's yet been answered. Dipodidae appears to have evolved in the Miocene from "a taxon at the sicistine/zapodine [viz, birch mouse/jumping mouse] level of evolutionary dental development" (Martin 1994, p. 99). Incidentally, Dipodidae is sometimes used for the clade that includes birch mice and jumping mice, as well as jerboas. However, most rodent workers seem to favour the use of the family-level name Zapodidae for birch mice and jumping mice, with Dipodidae restricted to jerboas proper. Dipodids + Zapodids = Dipodoidea. The name Dipodidae obviously comes from 'dipodes' meaning 'two-footed', the term apparently used for jerboas by Herodotus (writing some time around 430 B.C.).

On the subject of dipodid phylogeny I can't resist mentioning Krasnov & Shenbrot's (2002) study of co-evolution between fleas and jerboas. It's an interesting study in that they found no good correlation between flea phylogeny and jerboa phylogeny: the distribution of fleas on their jerboa hosts depends instead on ecological and geographical factors. You might argue that this is predictable, given that parasites like ticks, lice and fleas mostly switch between hosts that inhabit similar environments, rather than those that are closely related. Bed bugs *Cimex*, for example, are well known for parasitizing humans, but before this they were bird and bat parasites which "transferred their attentions to man when he began to live in caves and stayed with their new host when he moved away from the forests into other living quarters" (Andrews 1976, p. 162). Unfortunately for my research on *Euchoreutes*, it wasn't included in Krasnov & Shenbrot's (2002) study as it generally lacks fleas entirely!

Refs - -

- Andrews, M. L. A. 1976. *The Life That Lives on Man.* Faber & Faber, London.
- Dempsey, M. A. 1991. *Cranial foramina and relationships of dipodoid rodents.* Unpublished B. A. thesis, Baruch College of The City University of New York.
- Duplaix, N. & Simon, N. 1977. *World Guide to Mammals.* Octopus Books, London.
- Hanney, P. W. 1975. *Rodents: Their Lives and Habits.* David & Charles, Newton Abbot.
- Krasnov, B. R. & Shenbrot, G. I. 2002. Coevolutionary events in the history of association between jerboas (Rodentia: Dipodidae) and their flea parasites. *Israel Journal of Zoology* 48, 331-350.
- Martin, R. A. 1994. A preliminary review of dental evolution and paleogeography in the zapodid rodents, with emphasis on Pliocene and Pleistocene taxa. In Tomida, Y., Li, C. K. & Setoguchi, T. (eds) Rodent and Lagomorph Families of Asian Origins and Diversification. *National Science Museum Monographs* 8, 99-113.
- Nowak, R. M. 1999. *Walker's Mammals of the World, Sixth Edition, Volume II.* The Johns Hopkins Univesity Press (Baltimore and London).
- Shenbrot G.I., 1992. A cladistic approach to the analysis of phylogenetic relationships among dipodid rodents (Rodentia; Dipodoidea). *Archives of the Zoological Museum, Moscow State University* 29, 176-200.
- Rich, P. V. 1973. A mammalian convergence on the avian tarsometatarsus. *The Auk* 90, 676-677.

Chapter 19:
Giant furry pets of the Incas

If you've read Scott Weidensaul's excellent book *The Ghost With Trembling Wings* (2002), you'll recall the story of Louise Emmons and the giant Peruvian rodent she discovered. But before I get to that, let me say that *The Ghost With Trembling Wings* isn't about ghosts at all, but about the search for cryptic or supposedly extinct species. Think thylacines, British big cats, Ivory-billed woodpeckers, Cone-billed tanagers, the resurrection of the aurochs, Night parrots, Richard Meinertzhagen and the Indian forest owlet. It begins with Weidensaul's search for Semper's warbler *Leucopeza semperi*, an enigmatic parulid endemic to St. Lucia, discovered in 1870 and last seen alive in 1969 (although with a trickle of post-1969 sightings, some reliable and some not so reliable). If you're interested in the hunt for cryptic species and zoological field work and its history, it is mandatory that you obtain and read this inexpensive book.

Louise Emmons is a highly distinguished, experienced mammalogist who has worked on bats, tree shrews, cats big and small, and rodents, and is also the foremost expert on the mammals of the Neotropical rainforests (she wrote the only field guide to Neotropical rainforest mammals: Emmons 1999a). On 15th June 1997, while on an expedition to the northern Vilcabamba Range of Cusco, Peru, she was walking along a forest track when, lying dead on the track in front of her, she discovered a big dead rodent (Fig. 60). Pale grey, but handsomely patterned with a white nose and lips, and with a white blaze running along the top of its head, it was over 30 cm in head and body length, and with a

Fig. 60. The holotype specimen of Emmons' new chinchilla rat, shown as it was discovered.

tail over 20 cm long. Its broad feet, prominent and curved claws, large hallux, and palms and soles covered in small tubercles indicated that it was a tree-climbing species. A large bite wound on the neck indicated that it had recently been killed by a predator, probably a Long-tailed weasel *Mustela frenata*.

And it was entirely new: no one had ever recorded anything like it before. In her description of the new species, Emmons (1999b) named it *Cuscomys ashaninka* (meaning 'mouse from Cusco, of the Ashaninka people') and showed that it was a member of Abrocomidae. This is an entirely South American group previously known only from *Abrocoma* Waterhouse, 1837, members of which are sometimes called rat chinchillas, chinchilla rats or chinchilliones, and from the Miocene fossil *Protabrocoma* Kraglievich, 1927. *Abrocoma* is known from eight species (*A. bennetti*, *A. boliviensis*, *A. cinerea*, *A. vaccarum*, *A. uspallata*, *A. budini*, *A. famatina* and *A. schistacea*), among which *A. boliviensis* was only recognised in 1990 and *A. uspallata* in 2002 (Glanz & Anderson 1990, Braun & Mares 1996, 2002). Incidentally *A. bennetti* has 17 pairs of ribs – more than any other rodent. *Abrocoma* produces midden piles, and Pleistocene rodent middens from Chile have been identified by DNA analysis as having been produced by *Abrocoma* (Kuch *et al.* 2002).

Abrocomids are members of Hystricognathi, the rodent clade that includes Old World porcupines and the New World caviomorphs (New World porcupines, agoutis, pacas, cavies, pacaranas, capybaras, hutias, chinchillas, vizcachas and so on), and within this group they appear to be members of a clade that includes chinchillas and vizcachas.

So now there is a second extant abrocomid taxon, and it and *Abrocoma* are actually quite different. Species of the latter are specialized for life at high latitudes, and have short tails, a reduced hallux and inflated auditory bullae. They're entirely terrestrial, inhabiting burrows among rocks, and are therefore like chinchillas, and convergent on degus and pikas. While *Cuscomys* shares derived characters with *Abrocoma* not present in other rodents, it's larger, long-tailed and with features indicating a scansorial lifestyle. It's convergent with climbing murids, like the cloudrunners *Crateromys* and cloud rats *Phloeomys* of the Philippines and the giant tree rats *Mallomys* of New Guinea (Emmons 1999b), and its striking coloration is much like that of the White-faced tree rat *Echimys chrysurus* (a member of the echimyid, or spiny rat, family: echimyids are hysticognaths, as are abrocomids, but they apparently belong to the octodontid-hutia clade, not to the chinchilla-vizcacha clade (Sánchez-Villagra *et al.* 2003)).

Here's where this story becomes even cooler. During his 1912 Yale University-National Geographic expedition to the Inca ruins of Machu Picchu, Peru, George Eaton discovered that a number of different mammal species had been placed, in graves, alongside human bodies (Eaton 1916). They included familiar animals like dogs, llamas and guinea pigs, but also others that are far more obscure. Dwarf brockets were there (brockets *Mazama* are a group of small-bodied deer known from Mexico and South America), as were coro-coros (also called bamboo rats *Dactylomys*, coro-coros are arboreal members of Echimyidae), Mountain pacas *Cuniculus taczonowskii*, AND an abrocomid that Eaton recognized as a new species. He named it *Abrocoma oblativus* (incidentally, Eaton (1916) misidentified the Mountain pacas and thought that the Machu Picchu remains represented a new species that he called *Agouti thomasi*. Also of incidental interest is that the genus *Cuniculus* Brisson 1762 is the same animal as that more often called *Agouti* Lacépède 1799. The former name clearly has priority though).

Emmons's discovery of *Cuscomys ashaninka* allowed her to determine that the abrocomid in the Machu Picchu graves wasn't a species of *Abrocoma* as Eaton had thought, but a second member of *Cuscomys*, so it became renamed *C. oblativus*. Given what we now know of the life appearance of *Cuscomys*, it's likely that *C. oblativus* was similar: strikingly patterned, and overall quite cute and cuddly. There's the obvious implication here that Inca people were being buried with sacrified specimens of *Cuscomys* because they kept them as cuddly pets, though of course it's also possible that the animals were kept as food. If *C. ashaninka* has been cryptic enough to remain undiscovered until 1997, can we be absolutely sure that *C. oblativus* is really extinct? No. While the graves containing *C. oblativus* have been dated to 1450-1532 AD (Emmons 1999b), even today the region surrounding Macha Picchu is sparsely inhabited, remote, and with a substantial cover of pristine cloud forest. There just isn't any good reason why *C. oblativus* should have become extinct, so Emmons (1999b) suggested that it might still be extant, and awaiting rediscovery.

Here's another interesting thing. Of those mammals found in the Inca tombs, *Cuscomys* was unknown to modern scientists, in its living state, until 1997. The dwarf brocket present there turned out to belong to a new species that wasn't named until 1959 (when Hershkovitz named it *Mazama chunyi**), and the coro-coros and mountain pacas present in the tombs have also proved to be cryptic and elusive. So the Incas knew mammals that remained unknown to modern science until the late 20th century, and in fact knew them well enough to capture them frequently, and perhaps keep them in semi-domesticated state. As Emmons noted "Macha Picchu hunters were evidently skilled at capturing cloud forest mammals that are not readily taken by our current collecting methods" (1999b, p. 13). I know nothing of how Inca hunters tracked and caught the animals they did (nor do I have access to literature that might be informative on this subject), but it would be very interesting to know just how they were finding and catching these species. They must have had the most excellent, experienced field skills, and the most intimate knowledge of the species they were hunting.

Refs - -

- Braun, J. K. & Mares, M. A. 1996. Unusual morphological and behavioural traits in *Abrocoma* (Rodentia: Abrocomidae) from Argentina. *Journal of Mammalogy* 77, 891-897.
- Braun, J. K. & Mares, M. A. 2002. Systematics of the *Abrocoma cinerea* species

Mazama chunyi isn't the only recently-recognised brocket species. *M. permira*, a dwarf island-endemic from Isla San José off Panama, wasn't named until 1946, and *M. bororo* was named in 1996 after a specimen kept at Sao Paulo's Sorocaba Zoo demonstrated the distinctiveness of this taxon (Duarte & Gianonni 1996). What might be a new species was recently reported by Trolle & Emmons (2004) for a specimen photographed by a camera trap in 2003.

complex (Rodentia: Abrocomidae), with a description of a new species of *Abrocoma*. *Journal of Mammalogy* 83, 1-19.

- Duarte, J. M. B. & Gianonni, M. L. 1996. A new species of deer in Brazil (*Mazama bororo*). *Deer Specialist Group Newsletter* 13, 3.
- Eaton, G. F. 1916. The collection of osteological material from Machu Picchu. *Memoirs of the Connecticut Academy of Arts and Sciences* 5, 1-96.
- Emmons, L. H. 1999a. *Neotropical Rainforest Mammals: A Field Guide* (Second Edition). University of Chicago Press (Chicago & London).
- Emmons, L. H. 1999b. A new genus and species of abrocomid rodent from Peru (Rodentia: Abrocomidae). *American Museum Novitates* 3279, 1-14.
- Glanz, W. E. & Anderson, S. 1990. Notes on Bolivian mammals. 7. A new species of *Abrocoma* (Rodentia) and relationships of the Abrocomidae. *American Museum Novitates* 2991, 1-32.
- Kuch, M., Rohland, N., Betancourt, J. L., Latorre, C., Steppan, S. & Poinar, H. N. 2002. Molecular analysis of a 11,700 year-old rodent midden from the Atacama Desert, Chile. *Molecular Ecology* 11, 913-924.
- Sánchez-Villagra, M. R., Aguilera, O. & Horovitz, I. 2003. The anatomy of the world's largest extinct rodent. *Science* 301, 1708-1710.
- Trolle, M. & Emmons, L. H. 2004. A record of a dwarf brocket from Madre de Dios, Peru. *Deer Specialist Group Newsletter* 19, 2-5.
- Weidensaul, S. 2002. *The Ghost With Trembling Wings*. North Point Press (New York).

Chapter 20:

Osgood, Fuertes, and mice that swim and mice that wade

If you like the idea of being steeped in the lore of natural history research, then the literature on African amphibious murids provides rich pickings. Take the discovery of the obscure Ethiopian mouse *Nilopegamys plumbeus*, collected in 1927 by a field assistant of Wilfred H. Osgood on a tributary of the Blue Nile in north-eastern Ethiopia. Osgood (1875-1947) gained his reputation as an ornithologist and specialized during the 1890s in oology, but in 1897 he joined the then US Bureau of Economic Ornithology and Mammalogy (later to become the US Biological Survey) and embarked on significant collecting trips to California and Alaska. He later founded the Cooper Ornithological Club, and in 1909 joined the Field Museum of Natural History in Chicago. It was while based there that he made his best-known contributions: those published during the 1920s and 30s on the mammals of Africa (particularly Ethiopia) and South America (particularly Chile), and in particular on the rodents. In 1927, Osgood took part in an expedition jointly funded by the Field Museum and the *Chicago Daily News*.

In the field with Osgood was Louis Agassiz Fuertes (1874-1927), one of the most talented and revered of late 19th/early 20th century natural history artists (though he wasn't just an artist, as he also lectured). Predominantly interested in birds, Fuertes - like Osgood - had explored Alaska in the late 1890s but later traveled widely across the Americas and Africa. He illustrated countless books, magazines and museum murals. And, yes, he was named after the Harvard professor and naturalist Louis Agassiz.

During that 1927 field trip, it was Fuertes's job to draw the specimens obtained by Osgood's party. Presented with the new rodent later named *Nilopegamys*, the sketch Fuertes produced is

Fig. 61. *Nilopegamys plumbeus,* the Ethiopian water mouse. Photos of this species are not widely available. This painting is by Leon Pray and featured in Nowak (1999).

the only illustration that depicts a fresh specimen. It was also one of his last illustrations because, on returning home to the USA in August of that year, he was killed when a train struck his car at Potter's Crossing, Unadilla (New York). The illustrations in the car at the time – which included the *Nilopegamys* sketch – were thrown from the vehicle during the collision.

Osgood (1928) described *Nilopegamys* as an entirely new sort of murid for Africa: as an amphibious swimmer most like the South American fish-eating rats *Ichthyomys*. But *Nilopegamys* was clearly not as specialized for amphibious life as are the ichthyomyines, and furthermore Osgood's description was brief and without thorough comparisons to some other tropical African murids. Consequently it was suggested during the 1960s that *Nilopegamys* wasn't a distinct taxon, and that it should be sunk into synonymy with *Colomys goslingi*, the Velvet rat or African water rat. A long-limbed murid with an impressive array of whiskers, *Colomys* is amphibious and hunts for arthropods, worms and molluscs along stream and swamp edges. The consensus opinion became that Osgood and Fuertes had been incorrect about the validity of *Nilopegamys*. In time, it disappeared from the textbooks.

But it turns out that this decision was rash. Redescribing *Nilopegamys* in 1995, Julian Kerbis Peterhans and Bruce Patterson showed that *Nilopegamys* was clearly morphologically distinct from *Colomys*, and certainly worthy of generic recognition. While both genera are similar in their velvety fur and sharp demarcation between dark upperside and white underside, they differ in that *Nilopegamys* is larger, with broader feet that possess hairy margins, and with proportionally smaller ear pinnae. The two also differ in the arrangement of pads on their feet, in the number of roots their molars have, in the sizes of their foramen magnum, and in other details (Kerbis Peterhans & Patterson 1995). The features that distinguish *Nilopegamys* from *Colomys* suggest that it is more specialized for aquatic life than *Colomys* is. In essence, it seems to be evolving toward an ichthyomyine-like condition, and it certainly possesses several of the characters that Voss (1988) listed as being correlated with amphibious habits in murids (like dense and soft fur, enlarged hind feet, enlarged braincase and reduced visual and olfactory senses).

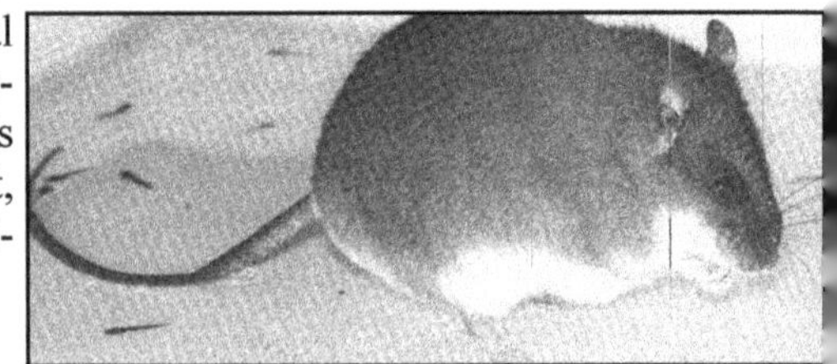

Fig. 62. The African water rat or Velvet rat *Colomys goslingi*. Photo by F. Dieterlen (from Nowak (1999)).

Nilopegamys and *Colomys* are both different from another tropical African form, *Malacomys*: the long-footed rats, big-eared swamp rats or long-eared marsh rats. Poorly known, but apparently ranging throughout most of central Africa, *Malacomys* looks like a mouse on stilts, at least when it's not crouching. Because these amphibious mice differ in so many of their anatomical details (and share few detailed, uniquely derived characters), Kerbis Peterhans & Patterson (1995) suggested that the similarities apparent between them are due to convergence. It would be nice, however, to test this by plugging them into a phylogeny. However, people are only just starting to work on parsimony-based murid phylogenies, and I'm not aware of any that incorporate *Nilopegamys* or *Colomys*. *Malacomys* at least appears to represent a distinct lineage within the so-called core murine clade, not too distant phylogenetically from *Mus* and *Apodemus* (Steppan *et al.* 2005), but there are suggestions that it is not monophyletic and that two distinct lineages may be included. McKenna & Bell (1997) listed *Nilopegamys* and *Colomys* adjacent to one another, but I suspect that this is an admission of ignorance more than anything else. The fantastically-named Congo forest mouse *Deomys ferrugineus* (sometimes called the Link mouse) is another similar long-footed form of uncertain phylogenetic position.

You might be surprised to hear that there are amphibious mice at all. But not only are there several amphibious African mice, there are in fact multiple murid lineages around the world whose members swim, wade, or forage in aquatic environments. South America is home to an endemic amphibious murid radiation, Ichthyomyini, that consists of five genera: three whose members are generally termed fish-eating rats (*Ichthyomys*, *Antomys* and *Neusticomys*), as well as *Rheomys* (the Central American water mice) and *Chibchanomys* (the Chibchan water mice). Only distantly related to these are *Holochilus* (the web-footed rats), *Nectomys* (the Neotropical water rats) and the recently discovered, poorly known *Lundomys* and *Amphinectomys*, all of which seem to be part of the rice rat [oryzomyine] group. Then there's *Scapteromys* (the 'rata acuatica'), which seems to be closest to the deeply weird *Kunsia* and *Bibymys* (all three genera have been united by some workers in a little clade termed Scapteromyini). Australasia has an assemblage of amphibious murids that includes *Hydromys* (the beaver rats) and *Crossomys* (the earless water rat), plus a number of genera where experts disagree as to whether the animals are actually amphibious or not. And there are others elsewhere in the world.

What makes the African taxa special is that, not only have they so far failed to become as well adapted for amphibious life as have murids elsewhere (like the ichthyomyines, or *Hydromys* or *Crossomys*), but they might also be doing something that murids elsewhere are not. For, while *Colomys*, *Malacomys* and *Deomys* are even less specialized for amphibious life than *Nilopegamys* is, they are specialized in one, peculiar way: recall that, earlier on, I characterized them as 'mice on stilts'. Noting that these genera possess particularly narrow, elongate feet, Kerbis Peterhans & Patterson (1995) showed that they formed a distinct cluster in terms of foot length : breadth, when compared with other murids. What might this mean? Unfortunately, very few published accounts discuss, describe or even mention the natural history and behaviour of these species, but a few do. While ichthyomyines and ichthyomyine-like murids are speedy swimmers that dart rapidly away underwater when threatened, Kerbis Peterhans & Patterson (1995) reported observations (made by Jonathan Kingdon and Ivan Sanderson) indicating that the stilt-legged mice really do use their long, narrow feet like stilts, wading around in shallow water. Such stilt-legged, wading mice seem unique to Africa.

Could it be that murids have evolved in this direction because there's something unique about tropical African waterways that has allowed them to specialize in this way? One thing does spring to mind: the presence of amphibious shrews and otter-shrews, all of which are, also, uniquely African. Living alongside *Colomys*, *Malacomys* and *Deomys* are the shrews *Ruwenzorisorex* and *Scutisorex*, both of which reportedly exploit aquatic environments (though, to be honest, you wouldn't know this from the literature). Murids are thought to have gotten into Africa relatively recently (about 6 million years ago), whereas the lipotyphlans have an African record going back as far as the Miocene.

Kerbis Peterhans & Patterson (1995) therefore suggested that "Prior or more successful exploitation of the 'swimmer' niche by lipotyphlans may have served to limit murid opportunities in this mode. Competition with lipotyphlans may also have driven the development of the 'wader' mode by central African forms" (p. 346). Such amphibious lipotyphlans are entirely absent from South America, and this might then explain why ichthyomyines have radiated so extensively. But if this is true, what about *Nilopegamys*, which (as I said earlier) is an African form apparently evolving toward an ichthyomyine-like condition? Well, it inhabits the Ethiopian plateau (and is in fact one of about 30 mammal species endemic to this region (Yalden & Largen 1992)), where there are no amphibious lipotyphlans.

Refs - -

- Kerbis Peterhans, J. C. & Patterson, B. D. 1995. The Ethiopian water mouse *Nilopegamys* Osgood, with comments on semi-aquatic adaptations in African Muridae. *Zoological Journal of the Linnean Society* 113, 329-349.
- McKenna, M. C. & Bell, S. K. 1997. *Classification of Mammals: Above the Species Level*. Columbia University Press, New York.
- Nowak, R. M. 1999. *Walker's Mammals of the World, Sixth Edition* (two volumes). The Johns Hopkins University Press (Baltimore and London).
- Osgood, W. H. 1928. A new genus of aquatic rodents from Abyssinia. *Field Museum of Natural History, Zoological Series* 12 (15), 185-189.
- Steppan, S. J., Adkins, R. M., Spinks, P. Q. & Hale, C. 2005. Multigene phylogeny of the Old World mice, Murinae, reveals distinct geographic lineages and the declining utility of mitochondrial genes compared to nuclear genes. *Molecular Phylogenetics and Evolution* 37, 370-388.
- Voss, R. S. 1988. Systematics and ecology of ichthyomyine rodents (Muroidea): patterns of morphological evolution in a small adaptive radiation. *Bulletin of the American Museum of Natural History* 188, 259-493.
- Yalden, D. W. & Largen, M. J. 1992. The endemic mammals of Ethiopia. *Mammal Review* 22, 115-150.

Chapter 21:
New, obscure, and nearly extinct rodents of tropical America, and... when fossils come alive

Though new rodents are described from all over the place (yes, even from North America and Europe*), I had a recollection of the greatest percentage coming from South America. And indeed there are quite a few (note that some of the following don't have common names), with a randomly-selected list of my favourites being...

-- the Candango or Brasilia burrowing mouse *Juscelinomys candango* Moojen, 1965, a semi-fossorial murid discovered in 1960, known from 9 specimens, last collected in 1990, and now possibly extinct (the Brazilian site where it was discovered was destroyed and built on).
-- Olrog's chaco mouse *Andalgalomys olrogi* Williams & Mares, 1978, an Argentinian sigmodontine murid.
-- *Abrawayaomys ruschii* Cunha & Cruz, 1979, a spiny Brazilian sigmodontine known from a handful of specimens.
- *Abrocoma boliviensis* Glanz & Anderson, 1990, a Bolivian chinchilla rat known from two specimens, one collected in 1926 and the other in 1955.
- *Amphinectomys savamis* Malygin *et al.*, 1994, an amphibious Peruvian murid known from a single specimen collected in 1991.
- *Pearsonomys annectans* Patterson, 1992, a semi-fossorial Chilean murid.
- *Microakodontomys transitorius* Hershkovitz, 1993, a Brazilian murid known from a single pecimen (collected in 1986).
- *Salinomys delicates* Braun & Mares, 1995, an Argentinian phyllotine sigmodontine with

North America recently yielded the Sonoma tree vole *Arborimys pomo* Johnson & George, 1991, and Europe the Bavarian pine vole *Microtus bavaricus* König, 1962. The latter species was thought extinct following its post-1962 isappearance, but was rediscovered in 2000.

proportionally long feet and large ears.
-- Roig's Chaco mouse *Andalgalomys roigi* Mares & Braun, 1996, another Argentinian sigmodontine.
-- Black or Koopman's tree porcupine *Coendou koopmani* Handley *et al.*, 1992, a Brazilian tree porcupine with short, dark fur. A similar form from Ecuador, differing in being speckled with white or yellow, was reported by Emmons (1999) and may be a new, as yet unnamed, species.
-- Orces fishing mouse *Chibchanomys orcesi* Jenkins & Barnett, 1997, an amphibious ichthyomine murid endemic to the Ecuadorian Parque Nacional Cajas.
-- *Cuscomys ashaninka* Emmons, 1999, a large Peruvian chinchilla rat (see Chapter 19) discovered in 1997.
-- *Akodon aliquantulus* Monica Díaz *et al.*, 1999, an Argentinian sigmodontine known from two specimens collected in 1993. The smallest member of its genus.
-- *Tapecomys primus* Anderson & Yates, 2000, a Bolivian phyllotine sigmodontine collected in 1991.
-- *Coendou ichillus* Voss & da Silva, 2001, an Ecuadorian tree porcupine first collected in 1936.
-- *Coendou roosmalenorum* Voss & da Silva, 2001, a small tree porcupine first collected in 1996 and named for Marc van Roosmalen and his son Tomas. Van Roosmalen is well known in South American mammalogy for the multiple new monkey species he has discovered.
-- *Abrocoma uspallata* Braun & Mares, 2002, an Argentinian chinchilla rat with larger ears and a longer tail than related species. Known only from a single specimen collected in 1995.
-- *Thomasomys ucucha* Voss, 2003, a sigmodontine (first collected in 1980) from the Cordillera Oriental of Ecuador.

There are many more. Note as usual, that the discovery date of a taxon is not necessarily the same as the date when it was first named, or recognised as new. Not all new rodents are South American: other discovery hotspots include Madagascar, Australia, New Guinea, and Borneo and elsewhere in SE Asia. Because rodents are typically small and inconspicuous it follows that a steady trickle of dull little mouse-type things should be continually discovered and described as new taxa, but it wouldn't be accurate to think that all new rodents are like this. Three of the animals listed above are tree porcupines, and these are all fairly big rodents with head and body lengths of about 30 cm. Abrocomids are also large, with head and body lengths typically exceeding 30 cm.

Discoveries of entirely new animals are very cool of course, but they're actually mundane and entirely ordinary. If you follow the literature it is very easy to become either overwhelmed or bored by the incredible number of new species that get described, even among tetrapods. Descriptions of new rodent, frog and lizard species appear routinely within technical journals – as in, a few every month. Perhaps slightly more interesting, and certainly more unusual, are those cases where species originally described as fossils have later turned out to be still extant. Such animals are often described as 'living fossils', but that's a bit silly given that virtually all extant species have a record going back many thousands of years at least, thus making their presence in the fossil record inevitable. Anyway, classic examples of this sort of thing include the following...

-- Goosebeak or Cuvier's beaked whale *Ziphius cavirostris*: described as a fossil in 1823 but realised in 1872 to be the same thing as beached specimens reported as early as 1826 but given different names.
-- Bush dog *Speothos venaticus* (Fig. 63): named as a fossil in 1839 [which explains why its generic name means 'cave wolf'], and first described in living form in 1843. The same person, Danish naturalist Peter Wilhelm Lund, described both the fossil and living animals, but failed to realise they were the same thing: he named the living animals *Icticyon*, and this name was used for *Speothos* until well into the 20th century.
-- False killer whale *Pseudorca crassidens*: described as a fossil in 1846 and described from modern-day strandings in 1862.
-- Mountain pygmy possum *Burramys parvus*: described from Pleistocene owl pellets in 1896 but found alive in a ski lodge in the Australian Alps in 1966.

Fig. 63. The peculiar Bush dog.

-- Chacoan peccary *Catagonus wagneri*: named as a fossil in 1930, and found alive in 1974.
-- Bulmer's fruit bat *Aproteles bulmerae*: described as a fossil in 1977 and reported from modern-day bones in 1980, then feared extinct, but since rediscovered alive.

Relatively little known is that the generic name for white-tailed and mule deers, *Odocoileus*, was originally coined for a fossil (a premolar found in a Pennsylvanian cave), and later transferred to the extant species when they and the tooth were found to belong to the same genus. Among rodents, there are, similarly, a few cases where fossil species have later been discovered extant, but because the animals concerned are obscure and poorly known, the relevant cases have gone under-reported...

-- A new fossil murid from Flores was described as *Floresomys naso* by Musser (1981). The generic name was preoccupied by a fossil Mexican sciuravid, so Musser *et al.* (1986) renamed this taxon *Paulamys naso*. A single live individual was reported in 1991 (Kitchener *et al.* 1991).
-- In 1887, Herluf Winge described multiple fossil murids from the Brazilian Lagoa Santa caves, and among them was a species he called *Scapteromys labiosus*. In 1980 this species, now referred to the crimson-nosed rat genus *Bibimys*, was reported to be extant within the same region (Voss & Myers 1991).

-- *Hesperomys simplex* was described from the Lagoa Santa caves by Winge in 1887, but also reported by him as occurring in modern-day owl pellets, and thus still extant. A Paraguayan murid named *Oryzomys wavrini* was described in 1921, and was shown by Voss & Myers (1991) to be the same thing as *Hesperomys simplex*, the name currently used for this taxon being *Pseudoryzomys simplex*. It's sometimes called the ratos-do-mato (Nowak 1999).
-- A living species from Uruguay and Brazil, described in 1955 as *Holochilus magnus*, was shown by Voss & Carleton (1993) to be the same thing as another Pleistocene fossil species named by Winge in 1887, *Hesperomys molitor*. Restudy of this murid showed that it was distinct from both *Holochilus* (the semiaquatic web-footed rats) and *Hesperomys* (nowadays synonymous with *Calomys*, the vesper mice) and thus deserving of its own genus, so today this species is called *Lundomys molitor*.

All of these 'prehistoric survivors' were known originally from Pleistocene or Holocene fossils, so their presence in modern times has only ever extended their geological range by a million years or so, at most (in some cases – such as that of *Paulamys naso* – by just a few thousand years).

Changing the subject somewhat, among modern-day mammals there are only two species whose discovery has extended the geological range of their clade by an amount of more than a few million years, and note that we're no longer talking about members of the same *species* being present across a longer span of time than originally thought. One of them has lately been in the news. The first is the so-called Monito del Monte or Colocolo *Dromiciops australis*, described in 1894 and classified as a didelphid. In 1955 however, Reig pointed out that *Dromiciops* was almost identical to *Microbiotherium* from the Miocene, and it is now widely agreed that *Dromiciops* is a living representative of Microbiotheriidae, a South American marsupial clade named in 1887 and with a fossil record that doesn't extend beyond the Lower Miocene. *Dromiciops* has no fossil record, so a ghost lineage of about 20 million years has to be invoked for the group. Incidentally, exactly how microbiotheriids fit into marsupial phylogeny is a hotly debated topic that would require a chapter all its own.

The second 'late survivor' brings us back to rodents: it's the Laotian kha-nyou *Laonastes aenigmamus*, described last year as representing an entirely new hystricognath lineage, the Laonastidae (Jenkins *et al.* 2005). But – how cool is this – Dawson *et al.* (2006) have shown that *Laonastes* is in fact a living representative of Diatomyidae, a group otherwise known only as fossils, and with a fossil record that doesn't extend beyond the Upper Miocene. So we now have to extrapolate a ghost lineage for diatomyids that extends from the Upper Miocene to the present: that's about 7-5 million years, so not that long, but... even so. As Dawson *et al.* (2006) note, late survivors that represent 'the reappearance of taxa after a lengthy hiatus in the fossil record' are termed 'Lazarus taxa', and *Dromiciops* and *Laonastes* can both be described this way.

And there's more to say on *Laonastes* too: it's exciting, not just in being a Lazarus taxon, but in being a specialised, highly cryptic member of a bizarre and specialised relict community. Tied to a specific unusual habitat, it is one of a suite of recently recognised species whose distribution may actually extend beyond Lao PDR.

Refs - -

- Min, M. S., Yang, S. Y., Bonett, R. M., Vieites, D. R., Brandon, R. A. & Wake, D. B. 2005. Discovery of the first Asian plethodontid salamander. *Nature* 435, 87-90.
- Dawson, M. R., Marivaux, L., Li, C.-k., Beard, K. C. & Metais, G. 2006. *Laonastes* and the "Lazarus effect" in Recent mammals. *Science* 311, 1456-332.
- Emmons, L. H. 1999. *Neotropical Rainforest Mammals: A Field Guide* (Second Edition). University of Chicago Press (Chicago & London).
- Jenkins, P. D., Kilpatrick, C. W., Robinson, M. F. & Timmins, R. J. 2005. Morphological and molecular investigations of a new family, genus and species of rodent (Mammalia: Rodentia: Hystricognatha) from Lao PDR. *Systematics and Biodiversity* 2, 419-454.
- Kitchener, D. L., How, R. A. & Maharadatunkamnsi. 1991. *Paulamys* sp. cf. *P. naso* (Musser, 1981) (Rodentia: Muridae) from Flores Island, Nasu Tenggara, Indonesia – description from a modern specimen and a consideration of its phylogenetic affinities. *Records of the Western Australian Museum* 15, 171-189.
- Musser, G. G. 1981. The giant rat of Flores and its relatives east of Borneo and Bali. *Bulletin of the American Museum of Natural History* 169, 67-176.
- Musser, G. G., van de Weerd, A. & Strasser, E. 1986. *Paulamys*, a replacement name for *Floresomys* Musser, 1981 (Muridae), and new material of that taxon from Flores, Indonesia. *American Museum Novitates* 2850, 1-10.
- Nowak, R. M. 1999. *Walker's Mammals of the World, Sixth Edition* (two volumes). The Johns Hopkins University Press (Baltimore and London).
- Voss, R. S. & Carleton, M. D. 1993. A new genus for *Hesperomys molitor* Winge and *Holochilus magnus* Hershkovitz (Mammalia, Muridae): with an analysis of its phylogenetic relationships. *American Museum Novitates* 3085, 1-39.
- Voss, R. S. & Myers, P. 1991. *Pseudoryzomys simplex* (Rodentia: Muridae) and the significance of Lund's collections from the caves of Lagoa Santa, Brazil. *Bulletin of the American Museum of Natural History* 206, 414-434.

Chapter 22:
Why azhdarchids were giant storks

People often talk of getting a 'culture shock' when they travel abroad. But in the world of zoological uber-nerdiness, you don't need to go abroad to experience a culture shock, you merely need to be an exposed to an idea that is shockingly alien and counter-intuitive. I still have fond memories of those days back in 1997 when I first visited the School of Earth & Environmental Sciences at the University of Portsmouth to be interviewed by the man who would later become my phd supervisor, Dave Martill. Of all the surprising things I was exposed to at the time, none was more striking and bizarre than the gigantic wall-mounted display on azhdarchid pterosaurs. Featuring a life-sized wing skeleton, some photos of a grinning German man (Dino Frey, guru of 'konstruction morphology') holding a bone, and a giant colour mural by John Sibbick, it's a pretty good exhibition, and eight years later it's still in the same place.

Fig. 64. An azhdarchid pterosaur imagined as a wading, stork-like predator of wetland environments. Image by Mark Witton.

Based on various of their morphological details, I had concluded that azhdarchids were most likely stork-like generalists that made their living by picking up assorted invertebrates and vertebrates, terrestrial and aquatic. Several other workers had also expressed a preference for this hypothesis. But Sibbick's giant colour mural depicted azhdarchids in an altogether different manner: they were shown wheeling above a vast expanse of ocean, gliding above the water surface and swooping down to grab fish. They were depicted as 'mega-skimmers', and Dave was later to explain to me why he and his colleagues favoured this skim-feeding hypothesis. I found it unconvincing and told him about the merits of the stork idea. He disagreed. We still disagree.

Today we have a new pterosaur worker in our research group, Mark Witton, and he's been looking at the morphology and palaeobiology of azhdarchids, among other things. At the risk of stealing his thunder (sorry Mark), I will say that he, also, is a supporter of the stork-like model. If you combine this with the recent work Dave and I have been doing on the non-azhdarchid azhdarchoids *Tupuxuara* and *Thalassodromeus* (the latter of which was claimed to be a skim-feeder), you can understand why the topics of skim feeding and azhdarchid lifestyles have become much discussed within our research group. Mark and I are obviously in happy agreement, but we still have to turn Dave around. I've wanted to write up my thoughts on this area for a while now, so here we go.

Interpreting azhdarchid palaeobiology

Since the announcement of its discovery in 1975, *Quetzalcoatlus* – the best known and best studied azhdarchid – has been imagined in several different ways. Initially it was interpreted as a vulture-like scavenger that soared over the Late Cretaceous landscape in search of dinosaur carcasses. Why was it interpreted this way? Well, apparently because *Quetzalcoatlus* was a big flying animal found in the same deposits as dinosaur bones (Lawson 1975). From a scientific perspective that's not exactly compelling.

A second hypothesis was proposed by Langston (1981): that azhdarchids fed on burrowing invertebrates by probing for them in the substrate (Fig. 65). This was also adopted by Wellnhofer (1991) (see p. 145, where Wellnhofer states "All this allows the possibililty that *Quetzalcoatlus* used its slender, pointed beak to search *in the ground* [my emphasis] for the molluscs and crabs that lived in the shallow pools of water"). Like Lawson's scavenging idea, this appeared to be based on

Fig. 65. Azhdarchids imagined as sediment-probing analogues of shorebirds. Image by Michael Rothman; first appeared in an article in *The New York Times*.

nothing more than circumstantial association of *Quetzalcoatlus* with other fossils, in this case invertebrate burrows. The same idea was endorsed by Lehman & Langston (1996), though only in an abstract. Lehman & Langston (1996) later came under extremely heavy fire from some workers, but the fact that Langston had actually proposed the probing idea first, and that Wellnhofer had agreed with it, seems to have been missed.

Describing the Asian azhdarchid *Azhdarcho*, Nesov (1984) was the first to propose a radically different hypothesis: that these pterosaurs might have behaved like *Rynchops*, the skimmers (Fig. 70). These charadriiform birds fly low over the water surface, trawling their unusual, laterally compressed lower jaws through the water, snapping up the fish and crustaceans that they make contact with. However, Nesov didn't base this idea on any features he observed among members of Azhdarchidae: he assumed that azhdarchids might have behaved this way, simply because other workers had earlier proposed a skimming lifestyle for *other* Cretaceous pterosaur groups. He wrote "If it is assumed that the Azhdarchinae could have flown like the Ornithocheirinae and Pteranodontinae – that is, like the Recent skimmers…" (p. 42). He went on to propose that azhdarchids might have been swimmers (that's not a typo) that captured aquatic prey, or that they might have "been able to hunt poorly flying vertebrates in the air". Neither of the latter ideas seems reasonable.

Paul (1987a, b) seems to have been the first to reject the scavenging hypothesis, at least in print, stating of *Quetzalcoatlus* that "Its slender, two-meter beak, with only thin bars around the external nares, is too weak for regular scavenging" (p. 20). Even better, he suggested a fourth possible lifestyle, proposing (Paul 1987a) that *Quetzalcoatlus* "probably patrolled water courses, like a three-meter-tall stork, picking up fish and small animals". I found Paul's argument moderately compelling when I first read it I-don't-know-how-many-years-ago, and I still find it moderately compelling today. Padian (1988) also rejected the scavenging idea and regarded azhdarchids as heron-like, noting that "Langston, who knows [*Quetzalcoatlus*] better than anyone, finds some suggestive resemblances to a heron or egret" (p. 64). Does this mean that Langston had given up on the probing idea? Noting the long, inflexible azhdarchid neck, Halstead (1989) wrote that

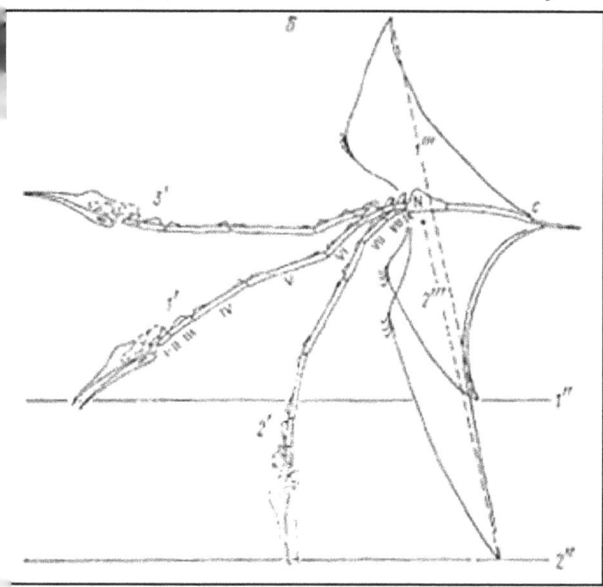

Fig. 66. Azhdarchid behaviour as imagined by Nesov (1984). This reconstruction shows (according to Nesov) how the long neck enabled the animal to reach down beyond the tips of its flapping wings.

"*Quetzalcoatlus* had a neck which could only move up and down and was specialized for dipping down into the water to snatch fish" (p. 160). This could perhaps be construed as agreement with the stork-like model.

In an article devoted exclusively to azhdarchid lifestyle, Iñaki Rodriguez Prieto (1998) argued against the scavenging hypothesis and agreed with Nesov's skimming idea. Morphological features were used to support the latter concept, but they were either erroneous or just incorrect. Prieto argued that the bill of *Quetzalcoatlus* was laterally compressed, that terrestrial abilities were very poor, and that the uropatagia would have made the hind limbs functionally similar to the forked tails seen in highly aerial birds like frigatebirds, swallows, and some kites. By implication, *Quetzalcoatlus* was deemed specialized for feeding on the wing, and it was on this basis that Prieto chose to follow the skimming hypothesis. All the features cited by Prieto are problematical, as we'll see later. Prieto's article was in Spanish and has been mostly overlooked.

Kellner & Langston (1996) also favoured the skimming hypothesis but did the same thing that Nesov did: they regarded it as 'at least plausible' for *Quetzalcoatlus* on the basis of the fact that it had been "previously advocated for *Rhamphorhynchus* ... and later assumed for many other pterosaurs, including the larger toothless pterosaurs" (p. 231). In other words, they didn't provide any supporting evidence at all, but merely elected to follow conclusions made for other, morphologically different pterosaur taxa (and, by the way, the concept of skim-feeding in rhamphorhynchids and those other pterosaurs isn't necessarily any more secure than it is for azhdarchids).

Unwin *et al.* (1997) noted that "unpublished functional studies and the circumstances of their preservation suggest that [azhdarchids] may have been piscivorous ... feeding from the water surface while on the wing and using the long neck as a 'fishing rod'" (p. 48). They cited an abstract and Unwin's unpublished thesis when discussing this area, but didn't elaborate as to why they favoured the skimming hypothesis. Martill (1997), while being harshly critical of Lehman & Langston's mud-probing idea (Fig. 67), also agreed with the skimming hypothesis, but, I would argue (sorry Dave), presented it as a just-so story, not as an evidence-led hypothesis. The presence of a 'highly streamlined skull' (p. 73) was used to support skimming behaviour, as was the long stiff neck. Martill *et al.* (1998) rejected both the scavenging and mud-probing hypotheses, and regarded azhdarchids as "aerial piscivores or planktivores" (p. 57). While noting the inflexibility of the neck, they didn't provide any supporting evidence for this hypothesis however.

Finally, Bennett (2001) noted that the femora and metatarsal V of *Quetzalcoatlus* were robust relative to those of other pterodactyloids, and that its feet were larger and more robust than those of ornithocheiroids. These observations led him to regard *Quetzalcoatlus* as "better suited to terrestrial locomotion than *Pteranodon*", and specialised for feeding on the ground. He wrote that "*Quetzalcoatlus* remains have been found far inland where they seem to have been heron- or stork-like in their ecology despite their great size" (p. 136).

So that's what's in the literature: four hypotheses (scavenger, skimmer, stork and mud-

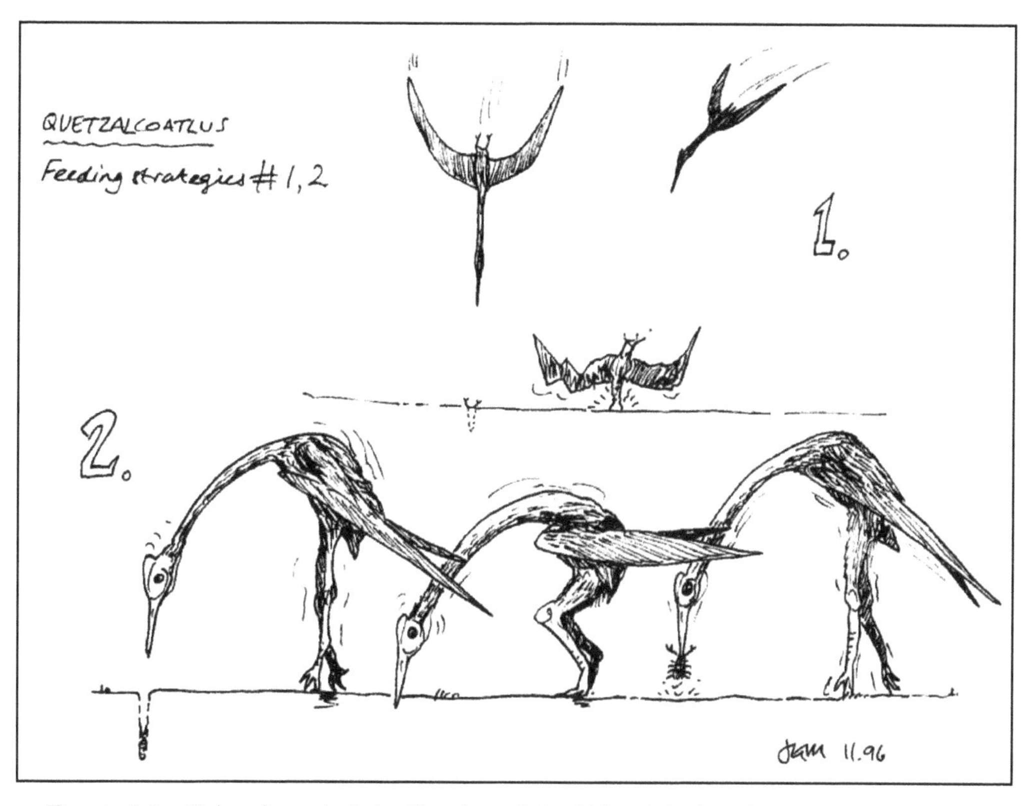

Fig. 67. Martill (1997) mocked the idea that azhdarchids might have been mud-probers, and published this cartoon (by John Martin).

prober). It's problematical that hardly any morphological features have ever been used to support them, and in fact some of them seem decidedly arm-wavy and intuitive, rather than evidence-led. So in the interests of introducing some hypothesis testing into this area, I've listed below all those hypotheses that I consider at least feasible, and have made predictions as to what evidence we would require in order to consider them reasonable. I do not consider Nesov's ideas that azhdarchids were swimming predators, or that they routinely captured flying vertebrates (presumably other pterosaurs and birds) likely, nor have I considered lifestyles that are obviously discordant with azhdarchid morphology (e.g., that they were deep divers, plunge divers, herbivores, aerial pirates or filter-feeders). I looked at as much literature as I could on form and function in bird bills, and also examined relevant specimens.

The hypotheses

Hypothesis 1: azhdarchids were vulture- or marabou-like scavengers, soaring over land and feeding from the carcasses of large terrestrial vertebrates. Predictions: proficient terrestrial abilities; bill and head capable of probing into body cavities; bill robust around narial openings; flexible neck; highly developed soaring or gliding skills. Note that hooked bill tips are not necessarily needed for this lifestyle, given that marabou storks and some corvids (e.g., Thick-billed raven *Corvus crassirostris*) routinely scavenge, yet have pointed bill tips. Witmer & Rose (1991) noted this.

Hypothesis 2: azhdarchids were mud-probers (like sandpipers), pushing their bills into sediment in search of burrowing prey. Predictions: proficient wading abilities (indicating by long, spreading toes); no need for highly developed soaring or gliding skills; skull should be specialized for probing, either with features allowing powerful gaping, with adaptations allowing rhynchokinesis, with a cross-section recalling that seen in mud-probing birds, and/or with well-developed tactile organs at the bill tip (e.g., Herbst corpuscles); neck should be reasonably flexible. Birds that probe sediment in search of prey have been shown to rely either on touch, on the detection of vibrations produced by the prey, or on the detection of pressure gradients surrounding hard objects (Gerritsen & Meijboom 1986, Piersma *et al.* 1998, Nebel *et al.* 2005). Pressure-sensitive organs termed Herbst corpuscles, embedded within pits on the premaxillary and dentary tips, are closely packed and particularly numerous in birds that probe sediments.

Hypothesis 3: azhdarchids were spear-fishers (like herons and anhingas), stalking fish in shallow water and spearing the body of the prey with sharp bill tips. Predictions: proficient wading abilities (indicating by long, spreading toes) or swimming abilities; sharply pointed, spear-like bill (Fig. 72); flexible neck that allows rapid darting of bill towards prey.

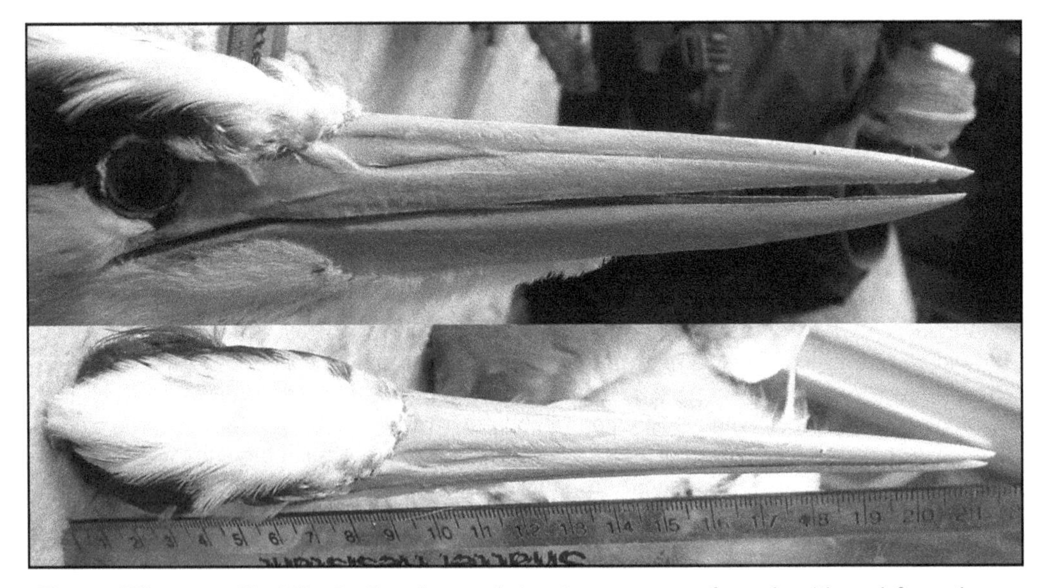

Fig. 68. The spear-like bill of a Grey heron *Ardea cinerea*, as seen from the side and from above.

Hypothesis 4: azhdarchids were skim-feeders, flying low over the water and trawling the lower jaw through the water. In contrast to that of many other birds, the feeding behaviour and cranial morphology of skimmers has been well described (Arthur 1921, Tomkins 1951, Bock 1960, 1964, Zusi 1962). Predictions: no need for proficient terrestrial abilities; highly skilled at fast, level flight; lower jaw laterally compressed and blade-like; streamlined bill; jaw joint, back of skull and neck able to withstand sudden jarring, with accessory articulation present between mandible and basicranium; upper jaw can be elevated relative to the basicranium (and is thus clear of the water surface during skimming); jaws capable of extremely rapid closure.

Hypothesis 5: azhdarchids were surface gleaners, or dippers, flying low over water and grasping food from the water surface (like albatrosses or frigate birds). Predictions: no need for proficient terrestrial abilities (hind limbs may even be strongly reduced); highly developed soaring or gliding skills; jaws elongate with down-curved tips; flexible neck allowing the animal to reach down and behind itself as it picks up food while flying over the water.

Hypothesis 6: azhdarchids were stork-like generalists, picking up assorted invertebrate and vertebrate prey from shallow water and/or terrestrial environments. Predictions: proficient terrestrial abilities; no need for highly developed soaring or gliding skills; bill elongate but lacking specializations (such as lateral or dorsoventral compression, keels, or hooked bill tips); neck flexibility not required as the neck only needs to bring the bill tips close to the ground; head-neck joint, at least, should be flexible.

The morphological evidence

Let's now see how the morphological data matches with these hypotheses and their predictions. Firstly, despite all those early claims making out that azhdarchids were like immense vultures, with tremendously elongate wings indicative of superb soaring or gliding skills, we now know that this was just not true. In fact, their legs were proportionally long, their wings were proportionally short compared to those of other large pterosaurs, and preserved wing membranes (which reveal that the brachiopatagium attached to the ankle) show that their wing membranes actually made their wings proportionally broad, and with low aspect ratios. As Frey *et al.* (2003) concluded, azhdarchids exhibited poor gliding performance compared with other large pterosaurs. Prieto's (1998) proposal that azhdarchid legs formed a pseudo 'forked tail' is nonsense given that the brachiopatagia incorporated the legs into the wings: they didn't trail behind the body as do a bird's tail feathers.

In terms of terrestrial abilities, we should note first that pterodactyloids in general were quite capable quadrupeds, and there is little reason to regard them as clumsy or helpless when grounded. Sure, they couldn't sprint at speeds equaling those of cursorial animals, but there is every indication that they were proficient walkers, more than capable of foraging quadrupedally on the ground or in shallow water. This is backed up by functional morphology, computer modelling, and evidence from trackways. With their proportionally short wings, long legs with robust femora, and large, robust feet (Bennett 2001), azhdarchids were likely to have been even better suited for terrestrial foraging than most other pterodactyloids. These lines of evidence suggest that azhdarchids were not specialized for a life on the wing (contra Prieto

1998): rather, they were better on the ground than were most other pterosaurs.

What does skull anatomy suggest? Good azhdarchid skulls are few and far between, with the best one being the incomplete rostrum described by Kellner & Langston (1996). In basic terms, the rostrum is shaped like a very long scalene triangle: it's deepest at the level of the nasoantorbital fenestra, but gradually tapers rostrally to a point. Some kind of bony crest is present over the caudal part of the nasoantorbital fenestra. Ignoring the crest, which living animals have a rostrum shaped like this? Storks, and not much else. Apparently the specimen described by Kellner & Langston (1996) is squashed flat, however, which makes it impossible to confirm whether the snout had the subrounded cross-sectional shape seen in storks. This is the spanner in the works, because if the skull is strongly compressed laterally, then the skull really isn't stork-like at all, but probably suited for some other, quite different mode of life. Like skimming.

But hold on: this isn't the only azhdarchid skull fragment known. Firstly, there's *Zhejiangopterus*. Again, we have a pointed, elongate, overall stork-like rostrum, but again the only figured specimen is apparently squashed flat, so it's not much use here. Aha, but there's *Azhdarcho*. Its rostrum fragments clearly belonged to a long, pointed (cough – stork-like – cough) rostrum that would have been subtriangular in cross-section, with the flat palate forming the triangle's base (Nesov 1984). An incomplete three-dimensional rostrum from Morocco, identified as '?Azhdarchidae', was described by Wellnhofer & Buffetaut (1999). It most certainly is not strongly compressed laterally, but is instead like a broad-based triangle in cross-section. Finally, a complete mandible is known for the Hungarian azhdarchid *Bakonydraco*. It's pretty odd, being pointed at its tip, slightly concave dorsally at the symphysis, and with a ventral mid-line ridge. The ridge is ventrally rounded, and not keel-like (Ősi *et al.* 2005). So, again, it's not laterally compressed.

Evaluation of the hypotheses

On the basis of all these features, how do the various hypotheses hold up?

Hypothesis 1 (the idea that azhdarchids were vulture- or marabou-like scavengers) doesn't stand up too well: while it's been all but rejected by some workers, note that they've only had scavenging raptors in mind, and haven't thought of comparing azhdarchids with marabou storks or scavenging corvids. In contrast to scavenging raptors, corvids and marabous, the azhdarchid rostrum does not appear to have been well braced around its openings (this is the naris in the birds, but the nasoantorbital fenestra in the azhdarchids), nor (with its bony dorsal crest) is the skull well suited for probing into body cavities, nor is the long, stiff neck in agreement with this lifestyle. Finally, while proficient terrestrial abilities were present, it does not seem that azhdarchids were specialized for long-distance soaring flight, as obligate scavengers are. I therefore feel that Hypothesis 1 can be rejected. In fact, even facultative scavenging like that present in marabous seems unlikely for azhdarchids, as (unlike marabous) their bills were weakly braced around the bony openings.

There are no cranial specializations consistent with Hypothesis 2 (mud-probing). Azhdarchid bill tips most certainly lack the sensory pits that house Herbst corpuscles in birds, though whether these would be present in a mud-probing pterosaur anyway is a good question. Regardless, the long, stiff azhdarchid neck doesn't match what is predicted for mud-probers either, and this hypothesis must also be rejected. Hypothesis 3 (spear-fishing) can also be rejected given that it is hard to imagine how the long, stiff neck could permit rapid lunging, stabbing and/or grabbing, plus the bill tip morphology is not spear-like as it is in the birds that practice this lifestyle.

We next come to the most popular Hypothesis: number 3, the skimming one. Despite its popularity I have to say that this is weak and not supported by the morphological evidence. While azhdarchids may well have been skilled at fast, level flight (as is *Rynchops*), this lifestyle does not explain the probably proficient terrestrial abilities present in azhdarchids. More importantly, there is nothing in the azhdarchid skull showing that it was streamlined and laterally compressed as required for this hypothesis, nor is there any indication that the jaw joint or back of the skull was built to withstand jarring impacts, nor that the upper jaw could be elevated relative to water level, nor that the jaws were capable of rapid closure, as is the case in *Rynchops* (Tomkins 1951, Bock 1960, 1964, Zusi 1962). The predictions are not fulfilled, so the skimming hypothesis is rejected.

Or, at least, the hypothesis that azhdarchids were *obligate* skimmers is rejected. Tomkins (1963) wrote of his surprise on learning that Royal terns *Thalasseus maximus* and Caspian terns *Hydroprogne caspia* are both capable of skimming behaviour, even though they lack the many unusual features present in *Rynchops*. Might azhdarchids, also, have been facultative skimmers? I would say that we can't rule it out, but (1) there's no evidence in its support and it's therefore nothing more than a speculation, and (2) it's still less well supported than other hypotheses.

We can also reject the rather similar Hypothesis 5: that azhdarchids were albatross-like surface gleaners, picking up prey from the water surface. The birds that do this are specialized for gliding and lack proficient terrestrial abilities, they have to have a flexible neck as they need to reach down and behind themselves as they pick up prey from the water surface, and they all have down-curved bill tips, presumably to aid in grabbing prey.

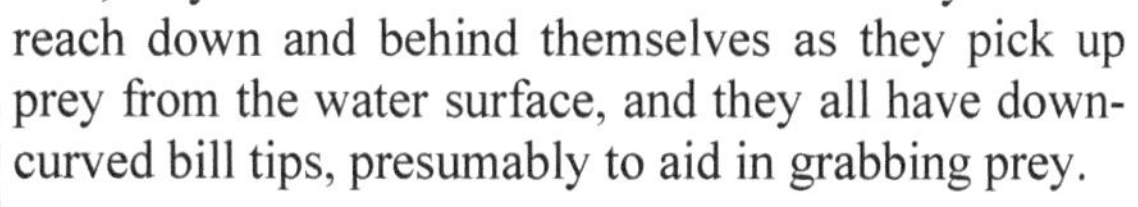

Finally, there's Hypothesis 6: that azhdarchids were stork-like generalists, picking up assorted invertebrate and vertebrate prey from shallow water and/or terrestrial environments. So far as I can tell, this is the only hypothesis where all of the predictions are

Fig. 69. Black skimmer *Rynchops nigra*, doing what it's known for. A good argument can be made that azhdarchids did not and could not do this.

met. Azhdarchids have the proficient terrestrial abilities required for a stork-like lifestyle, and lack features indicating a dedicated aerial lifestyle. Their jaws are elongate but lack the specializations present in skimmers, mud-probers or surface gleaners, and their long, straight neck vertebrae indicate that they could only raise and lower the neck vertically. That's ok for picking up animals from the ground and/or the water, but not much else. I therefore find Hypothesis 6 to be the only one that matches the evidence.

Conclusions

So having completed this little exercise I still regard the skimming hypothesis as poorly founded and problematic, and I remain very much in favour of the stork hypothesis. It should be noted that azhdarchids lack the specializations seen in some stork taxa. *Mycteria* (wood storks), for example, has a gently down-curved bill, a particularly dense array of Herbst corpuscles, and muscles that allow the jaws to be closed within 25 milliseconds (one of the fastest reflexes among vertebrates). These features are used by the birds as they search – using touch alone – for submerged prey (Hancock 1985). *Anastomus*, the Open-billed stork, has scopate tomial edges (meaning that it possesses tiny brush-like structures along the margins of its bill) and upper and lower jaws that bow away from each other, meaning that their edges never meet. These are apparently specializations that assist in the holding of hard-shelled prey (Gosner 1993). Rather, azhdarchids seem most like the most generalized storks, such as the *Ciconia* species. These eat everything from large insects, to frogs, fish, small crocodilians and mammals, and they patrol marshy areas and flooded meadows as well as dry grasslands for such prey. In fact they can make a living just about everywhere, and if you wanted to you could draw another parallel with azhdarchids here.

More research on this area is needed, but having said that I realize I've pretty much just written the better part of a paper on the subject. At some stage I'll re-vamp it for publication... perhaps with Mark as co-author.

Refs - -

- Arthur, S. C. 1921. The feeding habits of the Black skimmer. *The Auk* 38, 566-574.
- Bennett, S. C. 2001. The osteology and functional morphology of the Late Cretaceous pterosaur *Pteranodon*. Part II. Size and functional morphology. *Palaeontographica Abteilung A* 260, 113-153.
- Bock, W. J. 1960. Secondary articulation of the avian mandible. *The Auk* 77, 19-55.
- Bock, W. J. 1964. Kinetics of the avian skull. *Journal of Morphology* 114, 1-42.
- Frey, E., Buchy, M.-C. & Martill, D. M. 2003. Middle- and bottom-decker Cretaceous pterosaurs: unique designs in active flying vertebrates. In Buffetaut, E. & Mazin, J.-M. (eds) *Evolution and Palaeobiology of Pterosaurs. Geological Society Special Publication 217*. The Geological Society of London, pp. 267-274.
- Gerritsen, A. F. C. & Meijboom, A. 1986. The role of touch in prey density estimation by *Calidris alba*. *Netherlands Journal of Zoology* 36, 530-562.
- Gosner, K. L. 1993. Scopate tomia: an adaptation for handling hard-shelled prey? *Wil-*

son Bulletin 105, 316-324.
- Halstead, B. 1989. *Dinosaurs and Prehistoric Life*. Wm Collins Sons & Co., Glasgow.
- Hancock, J. 1985. Storks and spoonbills. In Perrins, C. M. & Middleton, A. L. A. (eds) *The Encyclopedia of Birds*. Guild Publishing (London), pp. 72-81.
- Kellner, A. W. A. & Langston, W. 1996. Cranial remains of *Quetzalcoatlus* (Pterosauria, Azhdarchidae) from Late Cretaceous sediments of Big Bend National Park, Texas. *Journal of Vertebrate Paleontology* 16, 222-231.
- Langston, W. 1981. Pterosaurs. *Scientific American* 244 (2), 92-102.
- Lawson, D. A. 1975. Pterosaur from the latest Cretaceous of west Texas: discovery of the largest flying creature. *Science* 187, 947-948.
- Lehman, T. M. & Langston, W. 1996. Habitat and behavior of *Quetzalcoatlus*: paleoenvironmental reconstruction of the Javelina Formation (Upper Cretaceous), Big Bend National Park, Texas. *Journal of Vertebrate Paleontology* 16 (Suppl. 3), 48A.
- Nebel, S., Jackson, D. L. & Elner, R. W. 2005. Functional association of bill morphology and foraging behaviour in calidrid sandpipers. *Animal Biology* 55, 235-243.
- Nesov, L. A. 1984. Upper Cretaceous pterosaurs and birds from central Asia. *Paleontology Journal* 1984 (1), 38-49.
- Ősi, A., Weishampel, D. B. & Jianu, C. M. 2005. First evidence of azhdarchid pterosaurs from the Late Cretaceous of Hungary. *Acta Palaeontologica Polonica* 50, 777-787.
- Padian, K. 1988. The flight of pterosaurs. *Natural History* 97 (12), 58-65.
- Paul, G. S. 1987a. The science and art of restoring the life appearance of dinosaurs and their relatives - a rigorous how-to guide. In Czerkas, S. J. & Olson, E. C. (eds) *Dinosaurs Past and Present Vol. II*. Natural History Museum of Los Angeles County/ University of Washington Press (Seattle and London), pp. 4-49.
- Paul, G. S. 1987b. Pterodactyl habits - real and radio-controlled. *Nature* 328, 481.
- Piersma, T., van Aelst, R., Kurk, K., Berkhoudt, H. & Maas, L. R. M. 1998. A new pressure sensory mechanism for prey detection in birds: the use of principles of seabed dynamics? *Proceedings of the Royal Society of London B* 265, 1377-1383.
- Tomkins, I. T. 1951. Method of feeding of the Black skimmer *Rynchops nigra*. *The Auk* 68, 236-239.
- Tomkins, I. T. 1963. Skimmer-like behavior in the Royal and Caspian terns. *The Auk* 80, 549.
- Unwin, D. M., Bakhurina, N. N., Lockley, M. G., Manabe, M. & Lu, J. 1997. Pterosaurs from Asia. *Journal of the Paleontological Society of Korea, Special Publication* 2, 43-65.
- Wellnhofer, P. & Buffetaut, E. 1999. Pterosaur remains from the Cretaceous of Morocco. *Paläontologische Zeitschrift* 73, 133-142.
- Witmer, L. M. & Rose, K. D. 1991. Biomechanics of the jaw apparatus of the gigantic Eocene bird *Diatryma*: implications for diet and mode of life. *Paleobiology* 17, 95-120
- Zusi, R. 1962. Structural adaptations of the head and neck in the black skimmer *Rynchops nigra* Linneaus. *Publications of the Nuttall Ornithological Club* 3, 1-101.

Chapter 23:
Hunting Green lizards in Dorset: new aliens or old natives?

If you like amphibians and reptiles (or, if you want to be as zoologically specific as possible: lissamphibians and non-avian reptiles), Britain is a pretty crappy place to live, with only a handful of natives. But strange things are happening: the diversity of our herpetofauna is being boosted by newly-appearing aliens, and some of our 'long-established aliens' are turning out to be genuine natives. Yesterday I spent the better part of the day in the field looking for lizards, and of the three species I saw, two are not natives. Probably.

Fig. 70. A male and female Western green lizard (the male is on the left). A Wall lizard *Podarcis muralis* joins the couple in the background.

The trip (organized by the Southampton Natural History Society) was to Southbourne and Boscombe Cliffs, Dorset, and the main purpose was to see Western green lizards *Lacerta bilineata* (this species was until recently included in *L. viridis* but a 1997 revision separated *L. bilineata* and *L. viridis* as separate species (Amann *et al.* 1997).

This complicates things, as it means that pre-1997 authors are not referring to our modern concept of *L. viridis* when they use this name. Indeed, much of the text you're about to read may be inaccurate as a result of this). This is a large lacertid – c. 130 mm snout-to-vent, and c. 300 mm in total length – that inhabits much of continental Europe except for the north-east: it's also devoid from much of Spain and Portugal where it's replaced by its relatives the Ocellated lizard *Timon lepidus* and Schreiber's green lizard *L. schreiberi*. Vivid green and decorated with fine dark stippling, and with a yellowish throat that turns blue in breeding males, it's an impressive looking, distinctive animal. It prefers to live in sunny places where there is bushy vegetation, and it even climbs quite high in bushes and shrubs to forage or bask (Arnold *et al.* 1992).

Green lizards don't appear to be native to the British Isles today, with the exception of Jersey where they inhabit south-facing coastal heaths and dunes. They also occur on neighbouring Guernsey, but here they've apparently been introduced from Jersey (Beebee & Griffiths 2000). It follows, then, that the announced discovery of a green lizard colony on the well-vegetated Dorset cliffs of Bournemouth in August 2003 came as quite a surprise. These were the lizards we had gone to see: we got to within very close range of a lone male who was sunning himself among some bramble, and we got some particularly good photographs of this individual. They're not the only lizards here: there are also Viviparous or Common lizards *Zootoca vivipara* and Wall lizards *Podarcis muralis* (both of which we saw too by the way. Damn we're good). The former is definitely a native, while the latter is almost certainly not. What are these aliens doing here, and are they aliens or not?

Green lizards are so pretty and charismatic that it comes as no surprise to learn that there have been numerous attempts to introduce them to Britain. Lever (1977) discussed introduction attempts that took place in Wales, Ireland, and Devon, the Isle of Wight, Surrey and Gloucestershire in England, with the earliest taking place in 1872. Some of these introduction attempts involved literally hundreds of animals, and must have been expensive undertakings. Numerous other releases are also known to have occurred, but, because there has never been any good indication of breeding occurring, Beebee & Griffiths (2000) noted that "on current evidence the species is not established in this country" (p. 208). Even so, lizards introduced to St Lawrence on the Isle of Wight in 1899 were apparently still there in 1936, so some introduced colonies have persisted for quite a while.

But there are also quite a few historical accounts that document Green lizards in places where, so far as we know, deliberate releases haven't occurred. As early as 1769, Gilbert White wrote of seeing 'Guernsey' lizards. It is generally thought that he had misidentified the definitely native Sand lizard *L. agilis* however (this is a stocky, short-limbed species which, although often green, is altogether easy to distinguish from *L. bilineata*, *L. viridis* and related species. It occurs as a native in Surrey, Dorset and Merseyside). Elsewhere, however, supposed green lizards were reported during the 1860s and 70s at Sidmouth and elsewhere in Devon, and during the 1900s in Dorset. One of the Devon records was made by a Mr John Wolley from Guernsey. He was apparently quite confident that the lizards he saw in Devon were exactly the same as the green lizards of Guernsey.

All of these reports were later dismissed as further misidentifications of Sand lizards, and while this might have been the case for some of the accounts, in others it appears unlikely, simply because the lizards described don't match this species. In some of the Dorset records it is noted that the lizards were nearly 300 mm long, for example, which is way too big for *L. agilis*. On the basis of these historical records, Jon Downes (1994) proposed that viable feral colonies of green lizards existed in Devon and Dorset, and that they were probably introduced from either France or the Channel Islands. He became disappointed that his idea was "ignored by the zoological establishment" and that "two famous zoologists (who shall remain nameless) told us that the theory was arrant nonsense. The paper was returned with a brusque letter from several zoological magazines and after a while we just gave up" (Downes 2003, p. 13). Given that, as noted above, relatively long-lived feral colonies had been reported earlier from the Isle of Wight, however, it would seem likely that Jon was right.

Fig. 71. Male Western green lizard. Photo by Neil Phillips.

The presence of what appears to be a viable, breeding colony, this time at Bournemouth, led Jon (Downes 2003) to write an article titled 'Told u so' [sic]. However, because there's no evidence that the Bournemouth colony is anything to do with the historical Devon and Dorset records discussed by Jon, it's not entirely satisfactory to claim that his contention has been vindicated. Then again, the fact that the Bournemouth colony is apparently viable and spreading (breeding is thought to have occurred) suggests that other colonies in southern England may well have been capable of this too. The Bournemouth colony was discovered by herpetologist Chris Gleed-Owen of the Herpetological Conservation Trust when he was on his way to work one day, and the colony is located just a few hundred metres away from Gleed-Owen's office (Gleed-Owen 2004).

Why are the lizards there? Gleed-Owen has suggested that they are dumped pets that have since bred. Could they have been introduced accidentally from the Channel Islands, or the continent, as Jon suggested for the other possible colonies? To answer this you'd need to know what sort of imports Bournemouth receives from abroad, and I haven't bothered to check that out. Finally, could they be late-surviving, hitherto-overlooked natives? This possibility has been inspired by the recent discovery that the (now extinct) British Pool frogs *Pelophylax lessonae* of Norfolk and Cambridgeshire were almost certainly natives, and not continental introductions as usually thought. It's also now being suggested that the European tree frog *Hyla arborea* colonies of the New Forest are also natives. Gleed-Owen regards the possibility of native status for the Bournemouth lizards as "unlikely, but not impossible".

I have to say that I think it's unlikely for two reasons: (1) the location is well known to naturalists and is regularly well explored. If the lizards occurred there long prior to 2003, it's difficult to believe that they would have been overlooked for so long. (2) All of Britain's native amphibians and reptiles have European ranges that extend far to the north-east, usually incorporating Scandinavia.

They are therefore relatively cold-tolerant, and this explains why they were able to colonise post-glacial Pleistocene Britain before the English Channel severed the continental connection about 9000 years ago. Conversely, green lizards are not animals of the north-east, which suggests that they wouldn't have been able to colonise post-glacial Britain during that brief critical phase, as our definite natives did. But the exciting possibility that they are natives can't be totally dismissed, and should be tested further.

Given that I've now written about British big cats (see Chapter 10) and introduced eagle owls (see Chapter 8), as well as green lizards, I will admit that I'm very interested in the alien species we have here, and the implications they have for ecology and diversity. Most people know that we already have, or have had, a lot of bizarre alien species in our islands (including bears, binturongs, wallabies, coypu, Edible and Garden dormouse, Midwife toads, American bullfrog, Ring-necked parakeets and so many others), but given the climatic changes that are occurring, it is likely that our fauna will change dramatically in future decades as old natives find life increasing difficult, and as new aliens find survival here easier.

Refs - -

- Amann, T., Rykena, S., Joger, U., Nettmann, H. K. & Veith, M. 1997. Zur artlichen Trennung von *Lacerta bilineata* Daudin, 1802 und *L. viridis* (Laurenti, 1768). *Salamandra* 33, 255-268.
- Arnold, E. N., Burton, J. A. & Ovenden, D. W. 1992. *Reptiles and Amphibians of Britain and Europe*. Collins, London.
- Beebee, T. & Griffiths, R. 2000. *Amphibians and Reptiles*. HarperCollins, London.

- Downes, J. 1994. Green lizards in Devon and Dorset? *Animals & Men* 2, 22-23.
- Downes, J. 2003. Told u so! *Animals & Men* 32, 12-13.
- Gleed-Owen, C. P. 2004. Green lizards and Wall lizards on Bournemouth Cliffs. *Herpetological Bulletin* 88, 3-7.
- Lever, C. 1977. *The Naturalized Animals of the British Isles*. Hutchinson & Co, London.

Chapter 24:

The hands of sauropods: horseshoes, spiky columns, stumps and banana shapes

I do quite a bit of consultancy work for companies that produce prehistoric animal books for children. In advising and assisting artists as often as I do, I find that they consistently screw up on the same things, every time. One of the biggest problem areas seems to be the hands and feet of sauropod dinosaurs – I reckon that every single artist whose work I've had to check has screwed up on these. By the way, the artists I'm talking about here lack palaeontological expertise or training: I'm **not** talking about your Luis Reys, Todd Marshalls and Mark Halletts, but rather about wildlife artists who find themselves being asked to illustrate dinosaurs.

Fig. 72. The hands of sauropods are bizarre. This is the right hand of the brachiosaur *Giraffatitan brancai*. Photo by Mike P. Taylor.

I also want to note that in no way is it the 'fault' of the artists concerned, given that (1) they've mostly based what they've done on the published work of those who have gone before them, and (2) while many of them have a history of working with palaeontologists, none of the experts they've been advised by before have bothered to tell them what they've been getting wrong. In fact I note that book-writing palaeontologists in general (you know who you are) rarely seem to bother providing their artists with any information, nor correct their mistakes, hence the incredible number of god-awful restorations that clutter the literature. Incidentally, this situation is getting worse as companies increasingly use CG images produced by people who seem to know nothing about animals, let alone fossil ones. But enough of that.

To save myself work in the future it occurred to me that it might be a good idea to write about sauropod hands and feet: that way, I can just direct interested parties to this chapter in future, and save myself the usual to-ing and fro-ing of notes, scans, and scribbled on diagrams. Several people have already produced overviews of sauropod hand and foot morphology, but unfortunately they're somewhat obscure and inaccessible to many. Greg Paul (1987) discussed in detail what sauropod hands and feet probably looked like, based on trackways and morphology, and provided a summarized version of the same information in his Japanese book *The Complete Illustrated Guide to Dinosaur Skeletons* (Paul 1996). Tracy Ford (1999) also published a guide to restoring sauropod hands and feet. Numerous technical studies describe or review sauropod hands and feet, with the most useful works being Christiansen (1997), Bonnan (2001, 2003, 2005) and Apesteguía (2005).

The basics

Let's note to begin with that sauropod hands and feet were only very superficially like those of elephants, and in fact in several details were fundamentally different. Sauropod hands, in particular, are unique in that the metacarpals did not spread out from the wrist as they usually do in tetrapods, but were arranged in a vertical column. My biggest peeve concerning restorations of sauropod hands and feet is that people seem unable to resist the temptation to illustrate multiple claws sticking out all over the place, or to depict hooves on all of the digits. As we'll see, the evidence is against both details. I know that many skeletal mounts give the animals multiple claws on the hands and/or feet, but that's because the mounts are inaccurate, outdated composites.

I should note here that the following discussion applies only to members of Eusauropoda, given that it now seems that basal sauropods outside of Eusauropoda lacked the distinctive hands and feet of the better known, fully graviportal sauropods.

The metacarpal colonnade

As noted, the eusauropod hand is formed from a columnar arcade of five vertically arranged metacarpals. If you were to look at the animal's hand from underneath (the plantar surface), the distal metacarpal tips would be seen to be arranged in a semi-circle, and the posterior surface of the hand would be concave. The hand was not backed by a pad (as McGowan (1991) wrongly stated), as it is in elephants. There is no doubt that this configuration was the case in life, as it's verified by numerous horseshoe-shaped hand tracks. In a new, as-yet-unpublished sauropod from the Lower Cretaceous of the Isle of Wight, it's been claimed that the first and fifth metacarpals virtually touch on the posterior surface of the hand, but this is unique so far as we know (this animal wouldn't have left horseshoe-shaped tracks, but subcircular ones… if the proposed interpretation is valid, and it might not be). An even more peculiar claim – in fact it's downright ridiculous – is that sauropods walked with their fingers curled under the distal ends of the metacarpals (Beaumont & Demathieu 1980). This is totally at odds with the morphological and trackway evidence and can be disregarded.

Not only did the metacarpals form this unique tubular structure, they and their digits seem to have been bound together to form a sort of pseudo-hoof: the digits didn't splay outwards from the metacarpus, but were bound together with them, nor did digits II to V possess hooves, claws or nails. This is supported by both anatomy and trackways, so distinct digits were almost certainly not visible in life.

Thumb claws, or lack of them

Claws were absent from all digits except for digit I (the thumb). This thumb claw varied in size, shape and orientation among the sauropod groups: it was particularly big in diplodocoids, where it was also laterally compressed and notably deep, and clearly separated from the rest of the metacarpus. In contrast, in brachiosaurids it was small, subtriangular in cross-section, and not separated from the rest of the metacarpus. In forms with large pollex claws (like diplodocoids), the anatomy of the penultimate phalanx and the distal end of the first metacarpal indicates that some flexion and extension of the claw was possible: in other words, these sauropods could both lift and lower their claws (albeit not by much compared to what was possible in other dinosaurs).

What did sauropods do with this lone thumb claw? Pretty much every conceivable function has been proposed, including fighting, digging, ripping foliage or tree trunk gripping. The claw's anatomy and range of motion led Upchurch (1994) to conclude that a trunk-gripping function was most likely, but if this is right then it seems odd that titanosaurs – the sauropods perhaps best suited for bipedality and rearing – were the ones that lost the claws.

That's right: even the thumb claw was not

Fig. 73. The right hand of an as-yet-undescribed new sauropod from Lourinhã, Portugal. Note the thumb claw pointing inwards on the inside surface of the hand.

present in all eusauropods. During the evolution of titanosauriform macronarians (the group that includes brachiosaurids and titanosaurs) the thumb claw was reduced in size and eventually lost altogether. We know that brachiosaurids such as *Brachiosaurus* possessed a short, small thumb claw, but Lower Cretaceous trackways apparently produced by other brachiosaurids indicate that the thumb claw was absent in these forms (it's been suggested that the thumb claw was still present in these forms, but that it was so small that it failed to leave an impression. I find this less likely than the idea that the claw really was absent). With the exception of the controversial Jurassic form *Janenschia* (and probably a few basal taxa from the Cretaceous), titanosaurs were all devoid of thumb claws.

Pads and spiky tubercles

But there's more – derived titanosaurs lacked not just claws, but finger bones too, and thus fingers. Their column-shaped hands were bizarre fingerless stumps, and they walked on the distal ends of the metacarpals, which is pretty odd to say the least. As described by Apesteguía (2005), the distal ends of the metacarpals in fingerless titanosaurs were wider and more rectangular than those of other sauropods, and with unusual sculpturing. The latter feature suggests that some kind of cushioning tissue encased the metatarsal ends, and a few titanosaurs preserve what appears to be some kind of soft tissue in this area.

Evidence for unusual soft-tissue structures on the sauropod metacarpus might have been present in other sauropods, according to recently described trackway evidence. A manus impression from the Upper Jurassic of Lourinhã in Portugal – probably produced by a brachiosaurid – preserves vertical score marks on its sides that seem to have been produced by rough tubercles on the metacarpal surface (Milàn *et al.* 2005). These authors proposed that at least some sauropods had spiky skin covering the distal end of the metacarpus (Fig. 74), though how widespread this was among sauropods we don't know.

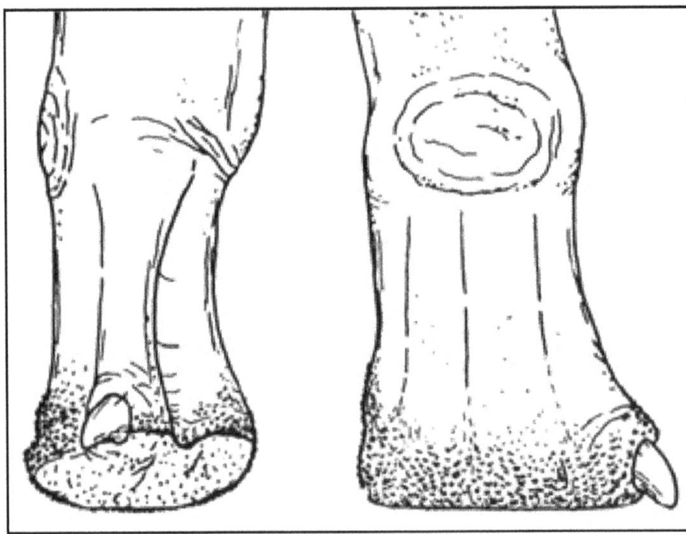

Fig. 74. The brachiosaur hand, as reconstructed by Milăn *et al.* (2005) with a covering of spiky tubercles. This diagram is based on an original drawing by Greg Paul.

Banana-shaped first metacarpals: why?

While all five metacarpals in most eusauropods were

more or less straight and parallel, this was not true of some titanosaurs. In these forms the first metacarpal was curved outwards at its distal end, and thus roughly banana-shaped. This is first seen in *Janenschia*, in which a thumb claw was present, but it's also the case in various other taxa, including *Andesaurus* and *Argyrosaurus*. Apesteguía (2005) made the intriguing suggestion that the bowed first metacarpal may first have evolved in claw-bearing basal titanosaurs in order to help support the claw, that it was then later retained when the claw was lost, and that it was later reversed (back to the straight condition) in derived lithostrotian titanosaurs.

This raises the question as to why a bowed first metacarpal was needed 'to help support the claw' however, given that other sauropods with thumb claws had straight, rather than bowed, first metacarpals. In *Janenschia*, a raised lip around the outer surface of the first metacarpal's distal end is very nearly in contact with the dorsoproximal part of the claw. This creates the impression that the metacarpal's distal end had evolved to help conduct stress along the curved ungual's dorsal margin. If that's true (let me emphasize that this is just an idea), it could mean that the thumb claw was used in a manner quite different from that of other sauropods. Maybe the claw tip was actually used for piercing something: presumably a substrate, or bark. Any better ideas?

Refs - -

- Apesteguía, S. 2005. Evolution of the titanosaur metacarpus. In Tidwell, V. & Carpenter, K. (eds) *Thunder-Lizards: The Sauropodomorph Dinosaurs*. Indiana University Press (Bloomington & Indianapolis), pp. 321-345.
- Beaumont, G. & Demathieu, G. 1980. Remarques sur les extremités antérieures des sauropodes (reptiles, saurischiens). *Compte Rendu des Séances de la Société de Physique et d'Histoire Naturelle de Genève* 15, 191-198.
- Bonnan, M. F. 2001. *The evolution and functional morphology of sauropod dinosaur locomotion*. Chapter 3: osteology of the forelimb. Unpublished phd thesis, Northern Illinois University.
- Bonnan, M. F. 2003. The evolution of manus shape in sauropod dinosaurs: implications for functional morphology, forelimb orientation, and phylogeny. *Journal of Vertebrate Paleontology* 23, 595-613.
- Bonnan, M. F. 2005. Pes anatomy in sauropod dinosaurs: implications for functional morphology, evolution, and phylogeny. In Tidwell, V. & Carpenter, K. (eds) *Thunder-Lizards: The Sauropodomorph Dinosaurs*. Indiana University Press (Bloomington & Indianapolis), pp. 346-380.
- Christiansen, P. 1997. Locomotion in sauropod dinosaurs. *Gaia* 14, 45-75.
- Ford, T. L. 1999. *How To Draw Dinosaurs, book 1*. T. L. Ford (privately published).
- McGowan, C. 1991. *Dinosaurs, Spitfires, & Sea Dragons*. Harvard University Press, Cambridge, Mass. & London.
- Milàn, J., Christiansen, P. & Mateus, O. 2005. A three-dimensionally preserved sauropod manus impression from the Upper Jurassic of Portugal: implications for sauropod manus shape and locomotor mechanics. *Kaupia* 14, 47-52.
- Paul, G. S. 1987. The science and art of restoring the life appearance of dinosaurs and

their relatives - a rigorous how-to guide. In Czerkas, S. J. & Olson, E. C. (eds) *Dinosaurs Past and Present Vol. II*. Natural History Museum of Los Angeles County/ University of Washington Press (Seattle and London), pp. 4-49.

- Paul, G. S. 1996. *The Complete Illustrated Guide to Dinosaur Skeletons*. Gakken.
- Upchurch, P. 1994. Manus claw function in sauropod dinosaurs. *Gaia* 10, 161-171.

Chapter 25:
The Cultured Ape, and Attenborough on gorillas

BBC4 just screened a whole night of documentaries on primates. I've sat up and watched such things as 'Natural History Night' and 'Dr Who Night' before: usually they're a con, the programmes fizzling out round about 10-30, but 'Primates Night' (err, if that's what it was called) wasn't so thrifty, keeping me in front of the TV until past 01-00 at least. And it was brilliant – the best assortment of TV programmes I've seen since, well, ever.

The first episode of the BBC series *Cousins* (presented by Charlotte Uhlenbroek) was shown: devoted to strepsirrhines, it included some great footage of aye-ayes, indris and sifakas. It's good, but I've seen it before (and got the book [Dunbar & Barrett 2000], but not the t-shirt). Two other documentaries were featured: they are among the best I've ever seen, and I really must get hold of copies. The first was essentially 'Frans de Waal's guide to cultural primatology'; it was Brian Leith's award-winning 2002 documentary *The Cultured Ape*.

Fronted by de Waal, and featuring Jane Goodall and a load of other primatologists whose names I've forgotten, *The Cultured Ape* concentrates on the many discoveries – well known to primatologists but still, it seems, alien to people at large – which show that humans are but one end of a behavioural and psychological spectrum, rather than an island separated from the rest of the animal kingdom. Concepts traditionally regarded as uniquely human, such as the development, maintenance and transmission of cultures, complex communication, and the use of tools, are of course now well recorded for chimps and other primates (e.g., Whiten *et al*. 1999, Byrne 2002), and a good case can be made that chimps and other non-humans also display less quantifiable traits such as guilt, deception, aesthetic enjoyment and hatred. When gorillas in zoos have looked after children who have fallen into their enclosures, are they not displaying altruism? Reportedly, mortally wounded chimps display behaviour that – if witnessed in a human – would be interpreted as pleading for their life.

Fig. 75. Jane Goodall, with friend.

On several occasions Goodall has explained the resistance she has received to her anthropomorphic interpretation of chimp behaviour after witnessing what seemed like jealousy, altruism, hatred and so on, and she explained how she learnt to couch these observations in a neutral language in order to get past reviewers. While I can understand that scientists want to avoid anthropomorphism, at the same time it seems unavoidable to conclude that individuals of at least some non-human species have personalities – surely everyone who's kept pets has experienced this, as Goodall stated on the programme – and can experience many/most of the same things that we can. Yet it seems that traditional ethology denies these as possible for other species, and in saying these things – the chimp enjoyed looking at the waterfall, the chimp felt guilty when it was caught stealing – one would be accused of being un-scientific. Some ethologists even argue that we shouldn't speak of non-human animals experiencing pain, given that we don't know that they are really experiencing the same sensation that we associate with that word. I've read some of the literature on this area, but I'm no ethologist, and I have no stake in this area. I'm just interested.

The other documentary – *Gorillas Revisited With David Attenborough* – was altogether different, but just as excellent. One of the most memorable and talked-about scenes from any TV documentary ever is Sir David Attenborough's 1978 encounter with wild Mountain gorillas *Gorilla beringei* in the Parc National des Volcans of Rwanda, broadcast in episode 12 ('A life in the trees') of the ground-breaking series *Life on Earth*, first broadcast in 1979. *Gorillas Revisited* covered the behind-the-scenes history of the 1978 filming, and what has happened to the Rwandan gorillas since. Attenborough was joined by *Life on Earth* producer John Sparks, cameraman Martin Saunders, and Ian Redmond, former assistant to Dian Fossey and now director of Global Great Ape Conservation. It was a fascinating story, tragically sad in places, uplifting in others.

Fig. 76. A Mountain gorilla mother and baby, photographed in Rwanda's Volcanoes National Park.

Most people – even those without a special interest in zoology – know the story of Dian Fossey and the gorillas she studied while at the Karisoke Research Center thanks to the 1988 film (and/or Fossey's 1983 book) *Gorillas in the Mist*. So it might come as no surprise to learn that the BBC team sought permission from Fossey to film 'her' gorillas, as they'd heard that this group had become habituated (= accustomed to humans). Somewhat surprisingly, Fossey gave enthusiastic approval, and urged Attenborough and colleagues to help promote the gorilla conservation work she, Redmond and her colleagues had initiated. At the time of the filming, Fossey was ill and still devastated by the recent killing of Digit, a young male gorilla who had been speared to death on New Year's Eve 1977. After the filming, the BBC team were shot at and arrested by the Rwandan army, who were under the impression that the film was being made in order to show what a bad job Rwanda was doing for gorilla conservation. The army also thought that the BBC had been filming Digit's body. They strip-searched Attenborough and confiscated the film cans, but the crew had cleverly swapped the labels on the cans, so the soldiers were confiscating unused film.

During the *Life on Earth* sequence one young gorilla clambers all over Attenborough and lies back on Attenborough's chest. From its behaviour you might assume that this gorilla was a plucky, bold and confident individual. Well, he was named Pablo, and today he is an adult silverback. Another individual who was a youngster when Attenborough encountered him, Titus, is today a silverback who leads a group of 59 animals: the biggest recorded gorilla group ever. Many of the gorillas named by Fossey are doing well today, and have become parents and grandparents, and Fossey will always be remembered for initiating one of the first long-term generation-level studies of a wild mammal population.

If you've seen *Gorillas in the Mist* you'll know that Fossey's work was actually inspired by her meeting with a palaeontologist, Louis S. B. Leakey: a great example of a very fruitful crossover between palaeontology and field biology. Prior to Fossey's work, Rosalie Osborn had studied Mountain gorillas. She also acknowledged Leakey's involvement in setting up her research (Osborn 1963), and Leakey is also acknowledged by Goodall as initially suggesting that she might study the chimps she eventually became so acquainted with (van Lawick-Goodall 1971). It's no secret that Leakey regarded women as better suited for observational fieldwork than men, apparently because he regarded women as more observant, and more patient.

Of course, things have not all been rosy in the Virungas. In 1968 half of the Parc National des Volcans was taken for pyrethrum cultivation (ironically, grown for use in Europe as an environmentally-friendly alternative to DDT) and

Fig. 77. A male Mountain gorilla.

there were plans, backed by the European Development Fund, to replace even more of the park with pyrethrum (Harcourt 1981). Poaching was a serious problem that Fossey and her successors have had to deal with (snares are set for hoofed mammals, but gorillas get caught and injured in them), and during the 1970s several gorillas were shot. Following the humanitarian crisis that engulfed the region following the civil war and resultant genocide of April-July 1994, rebels invaded the park and looted the homes and facilities, and murdered several of the people employed to keep the gorillas safe from poachers. Rebels also killed gorillas, and anti-poaching trackers, in 2001. And of course Fossey herself was murdered in 1985.

From a population that was thought to be around 600 in 1960, Rwandan Mountain gorillas had dwindled to an estimated low of 200 or so by 1980. Today, there are around 380 animals. That's better than it was, but still pitifully low.

Refs - -

- Byrne, R. W. 2002. Social and technical forms of primate intelligence. In de Waal, F. B. M. (ed) *Tree of Origin*. Harvard University Press (Cambridge, Mass. & London), pp. 145-172.
- Dunbar, R. & Barrett, L. 2000. *Cousins: Our Primate Relatives*. BBC Worldwide, London.
- Harcourt, A. H. 1981. Why save the mountain gorilla? *Wildlife* 23 (2), 22-26.
- Osborn, R. M. 1963. Observations on the behaviour of the Mountain gorilla. *Symposia of the Zoological Society of London* 10, 29-37.
- van Lawick-Goodall, J. 1971. *In the Shadow of Man*. William Collins Sons & Co, London.
- Whiten, A., Goodall, J., McGrew, W. C., Nishida, T., Reynolds, V., Sugiyama, Y., Tutin, C. E. G., Wrangham, R. W. & Boesch, C. 1999. Cultures in chimpanzees. *Nature* 399, 682-685.

Chapter 26:

At last, the Odedi revealed: the most mysterious bush warbler

In March 2006 Mary LeCroy and F. Keith Barker published their description of the Odedi *Cettia haddeni*, a bush warbler from Bougainville Island of the North Solomons Province in the SW Pacific (LeCroy & Barker 2006). Bougainville Island will be familiar to you if you've read about obscure and/or recently extinct birds, as it's famously home to the Moustached kingfisher *Actenoides bougainvillea* (a species which has only been seen or heard on a handful of occasions over the last few decades), among others.

Cettia, the bush warbler genus, is represented by 14 species, most of which live in SE Asia, but there are also several species that inhabit the islands of the SW Pacific. We have one member of the genus here in Europe: Cetti's warbler *Cettia cetti* (though it's not a European endemic, as it also occurs across Asia). People here don't ordinarily call it 'Cetti's bush warbler', but they do elsewhere in the world (in India for example, where it's but one of eight *Cettia* species). Cetti's warbler is a skulking bird that tends to stay hidden in river-bank foliage (Fig. 78), its distinctively explosive song giving its location away. Will and I used to go find them along overgrown canals when we lived in Gosport (well, I used to go find them. Will just came along for the ride).

Incidentally, it's a good example of a bird whose range has increased substantially within recent history. Early in the 20th century it was apparently restricted to the Mediterranean region, but it's been gradually spreading northward and today occurs as far north as Sweden (or at least that's what some of the books say: in the field guides it isn't shown as extending further north than southern Britain).

Bush warblers are particularly newsworthy right now (to my mind at any rate) given that the just-published oscine supertree of Jønsson & Fjeldså (2006) found *Cettia* to be diphyletic, with *C. cetti* grouping with the tesias (tesias are one of a number of poorly-known tropical passerines for which the common name is the same as the technical generic name. Other examples include prinias, yuhinas, minlas, newtonias, oxylabes, niltavas, liocichlas, apalis, camaropteras, eremomelas… the list goes on) and *Urosphena* (the stubtails) while the Japanese bush warbler *C. diphone* grouped with the Broad-billed flycatcher-warbler *Tickellia hodgsoni* and *Orthotomus* (the tailorbirds). Admittedly, the idea that Cetti's warbler might group with tesias and stubtails, both of which are radically short-tailed, seems odd, but then total evidence is the game.

Fig. 78. Cetti's warbler *Cettia cetti*. Photo by S. Jobling.

Results contrary to those of Jønsson & Fjeldså (2006) were found by LeCroy & Barker (2006): based on cytochrome *b* sequences, they found *Cettia* to be monophyletic, with *C. cetti* as a basal member of the clade and stubtails as the sister-group. The island endemic forms of the SW Pacific formed a subclade within *Cettia*, thereby supporting Orenstein & Pratt's (1983) contention that this was probably the case. It's been suggested that the island-endemic bush warblers descended from a wide-ranging colonizing ancestor that originated from a continental source. This is the standard stepwise dispersal model favoured for the evolution of island endemics: it's recently been shown that, remarkably, some continental passerine radiations descended from island passerine clades (Filardi & Moyle 2005), but there's no indication that this occurred in bush warblers.

Anyway, if island-endemic bush warblers have descended from a widespread colonizing ancestor, then the members of this clade are somewhat patchily distributed in the region, and their absence from some islands and island groups (such as New Guinea and the Bismarck Archipelago) seems odd. Maybe bush warblers *were* present on these islands and have since become excluded by more recently evolved species (the 'taxon cycle' model), or perhaps they've been made extinct by people. Such is supported by the fact that extinct bush warbler species have been reported from Tonga (Steadman 1993, 1995). Or… maybe bush warblers actually *do* inhabit some of these islands, but await discovery. This is possible given that other members of the group are recent discoveries: there's the Odedi of course, but also *C. carolinae* from Tanimbar in the Moluccas. It was only named in 1987 (Rozendaal 1987).

What makes the Odedi further interesting is that, prior to 2004, it was a mystery animal known only from ethnic reports and from its vocalisations, and in fact in the annals of obscure ornithology it has a relatively long and interesting history. In 1975 Jared Diamond noted his 1972 discovery of an unknown mountain-dwelling passerine on Bougainville Island, known to the local speakers of the Rotokas language as the kopipi, and to the Nasioi speakers as the ódedi. Diamond never saw the bird but did describe its thrush-like song (Diamond 1975). Over the following years several other ornithologists were to learn about and encounter the Odedi, including Don Hadden and Bruce Beehler, both well known specialists of the birds of the Solomons and surrounding Islands. Hadden heard the bird calling on many occasions between 1977 and 1980, usually during misty and/or wet weather, but he was never able to catch one. Beehler (who I'd say is best known for his work on birds-of-paradise) published a brief paper on the bird in 1983 and thought it most likely that it was a species of *Vitia* (Beehler 1983), a genus regarded nowadays as synonymous with *Cettia* (Orenstein & Pratt 1983).

Hadden was eventually able to obtain photos of the species, and he included one in his 2004 book *Birds and Bird Lore of Bougainville and the North Solomons* (Hadden 2004). Even better, at long last, he managed to obtain a specimen in January 2000, and subsequent specimens were procured later in the year and in 2001. As Hadden and others had predicted, the Odedi proved to be a rather plain, short-winged, chestnut-coloured bush warbler. Compared to other bush warblers of the SW Pacific, it is huge (fully 4 grams heavier than any other bush warbler), notably dark, with a wider bill, a more robust tarsus and longer toes.

So in view of Hadden's long quest for this species, culminating in its discovery, it is fitting that LeCroy & Barker (2006) have named it after him.

Given that the Odedi was known from its vocalizations and from ethnic reports prior to the procurement of the first specimen, does this mean that – pre-2000 – it was a cryptid? I would say yes. And if people who are regarded without question as card-carrying zoologists of the utterly ordinary type are actually out there chasing down ethnoknown species like the Odedi, are they actually cryptozoologists? That's less easy to answer, because cryptozoologists are more usually regarded as dedicated to the pursuit of cryptids alone, whereas Hadden and others don't search for species like the Odedi to the exclusion of all others. But I personally don't see any problems with the idea that 'ordinary' zoologists actually do engage in cryptozoological research at times, it's just that they tend not to identify the research as such. As it happens, there are actually a few small passerines known from observations or photos, but not (yet) from specimens.

Refs - -

- Beehler, B. 1983. Thoughts on an ornithological mystery from Bougainville Island, Papua New Guinea. *Emu* 83, 114-115.
- Diamond, J. 1975. Distributional ecology and habits of some Bougainville birds (Solomon Islands). *Condor* 77, 14-23.
- Filardi, C. E. & Moyle, R. G. 2005. Single origin of a pan-Pacific bird group and upstream colonization of Australasia. *Nature* 438, 216-219.
- Hadden, D. 2004. *Birds and Bird Lore of Bougainville and the North Solomons*. Dove Publications Pty (Alderley, Queensland).
- Jønsson, K. A. & Fjeldså, J. 2006. A phylogenetic supertree of oscine passerine birds (Aves: Passeri). *Zoologica Scriptca* 35, 149-186.
- LeCroy, M. & Barker, F. K. 2006. A new species of bush-warbler from Bougainville Island and a monophyletic origin of southwest Pacifc *Cettia. American Museum Novitates* 3511, 1-20.

- Orenstein, R. I. & Pratt, H. D. 1983. The relationships and evolution of the southwest Pacific warbler genera *Vitia* and *Psamathia* (Sylviinae). *Wilson Bulletin* 95, 184-198.
- Rozendaal, F. G. 1987. Description of a new species of bush warbler of the genus *Cettia* Bonaparte, 1834 (Aves: Sylviidae) from Yamdena, Tanimbar Islands, Indonesia. *Zoologische Mededelingen* 61, 177-202.
- Steadman, D. W. 1993. Biogeography of Tongan birds before and after human impact. *Proceedings of the National Academy of Sciences, USA* 90, 818-822.
- Steadman, D. W. 1995. Prehistoric extinctions of Pacific Island birds: biodiversity meets zooarchaeology. *Science* 267, 1123-1131.

Chapter 27:
When animals die in trees

I like living animals as much as the next person, but I also can't help but find dead ones just as interesting. Corpses, decomposition, burial... all of it I find fascinating. Most palaeontologists do, and in fact there are whole books written about what happens to animals after they die. This area of study is termed taphonomy, and part of my palaeontological training has involved taphonomic interpretation of fossils as well as actualistic work on modern bodies and bones. I'll be writing about some of this actualistic work in the near future.

Many tetrapods that climb or rest on, in, or under branches, cave roofs and so on have evolved structural adaptations that allow their digits to grip automatically, and with only a minimal expenditure of energy. If we humans want to grasp firmly to a branch overhead, we have to use a lot of muscular effort to keep our digits flexed. But bats, passerines and members of other groups have evolved tendon locking mechanisms. If I start talking about these adaptations, I'll easily add another 1000 words to this chapter, so I'm going to avoid doing that and will direct you to the literature instead. For the tendon locking mechanisms of bat feet see Schutt (1993) and Quinn & Baumel (1993), for bats as well as dermopterans see Simmons & Quinn (1994), for rodents see Haffner (1996) and for birds see Quinn & Baumel (1990).

But here's a thought. If your digits automatically secure a purchase to a branch, or the roof of a cave, then what happens when you die? Answer: your corpse stays there, dangling for all eternity.

Some years ago I spent time collecting literature on accidental deaths. A Cape buffalo collides with a tree and dies of a broken neck, a giraffe gets hit by lightning, a sparrowhawk gets stuck in foliage while pursuing a passerine... that sort of thing. Numerous accounts describe death by choking (particularly among snakes, predatory lizards, aquatic birds and raptors). Poisoning is not uncommonly reported (usually where inexperienced young herbivores eat toxic plants). But what of deaths in trees? To start with, there are the animals that die after getting their necks caught in forked branches, as has been recorded for both giraffes and deer. Honestly. Then there are animals that get tangled up in thorny or spiky vegetation. This is particularly well reported for bats, with Hill & Smith (1984) discussing cases of bats dying after being entangled in burdock burs, after getting impaled on the spines of cacti, desert shrubs and rose thorns, and after getting strangled by Spanish moss fibres.

But I haven't just been collecting reports of strange deaths, I've been collecting corpses too, and lots of them. A fox that died on a bed of straw in a stable. Pressed toads. Road-killed deer. Finches, wrens and bats that were killed by cats. A shrew that died on a railway platform. Moles, voles, newts, frogs, snakes, pigeons, thrushes.

Fig. 79. A European robin *Erithacus rubecula* corpse, still dangling from its perch.

Two of my favourite corpses died in trees (or, at least, big shrubs). The first is a European robin *Erithacus rubecula.* Already highly decomposed, and consisting only of a feathered skeleton with some ligaments, it was discovered within a 2 m high privet hedge, and within the hedge the bird was about 1.2 m off the ground. Not only was its little corpse still stuck in the hedge where it died, one of its feet was still curled tightly around a branch, and in fact the bird was hanging upside down by this one foot. My interpretation is that the bird died while sleeping in the hedge, and that after death its foot remained clenched around the branch. I carefully removed the bird, and the branch it was attached to, and I still own them today (Fig. 79). As you can see, the corpse still hangs from the branch by a single foot.

Given that so many bird species hide, nest and roost in trees, we should expect them to die in trees quite often. Small passerines even secrete themselves deep into tangled vegetation and into cracks and fissures in bark. If they then die in their sleep, they're just going to stay stuck there. While I've checked and collected quite a lot of literature on mortality in birds and other tetrapods, I have yet to read of any other instances of birds being found dead in trees. Surely other instances of this have been reported. Anyone? Squirrels are often found dead in their dreys (Gurnell 1987).

My second favourite corpse is a Grey squirrel *Sciurus carolinensis*, and it also died in a tree. But this case is a little odd, as the corpse was discovered wedged in thin branches about 3 m off the ground. Stuck up there, it began to decompose. Its skin ruptured. Its guts fell part-way out of its abdomen. But the soft tissues didn't all rot away: they became wind-dried, and the corpse (with the hanging guts) became mummified. Today, the corpse has a large U-shaped concavity on its ventral surface, caused by a supporting branch that was under the body while the rest of it sagged down toward the ground (Fig. 80). Ordinarily, any squirrel that dies in a tree falls to the ground. Gurnell (1987) reported on a 1985 mortality event in England where squirrels were mysterious dropping dead, and "one was actually seen dying and falling off a branch" (p. 135).

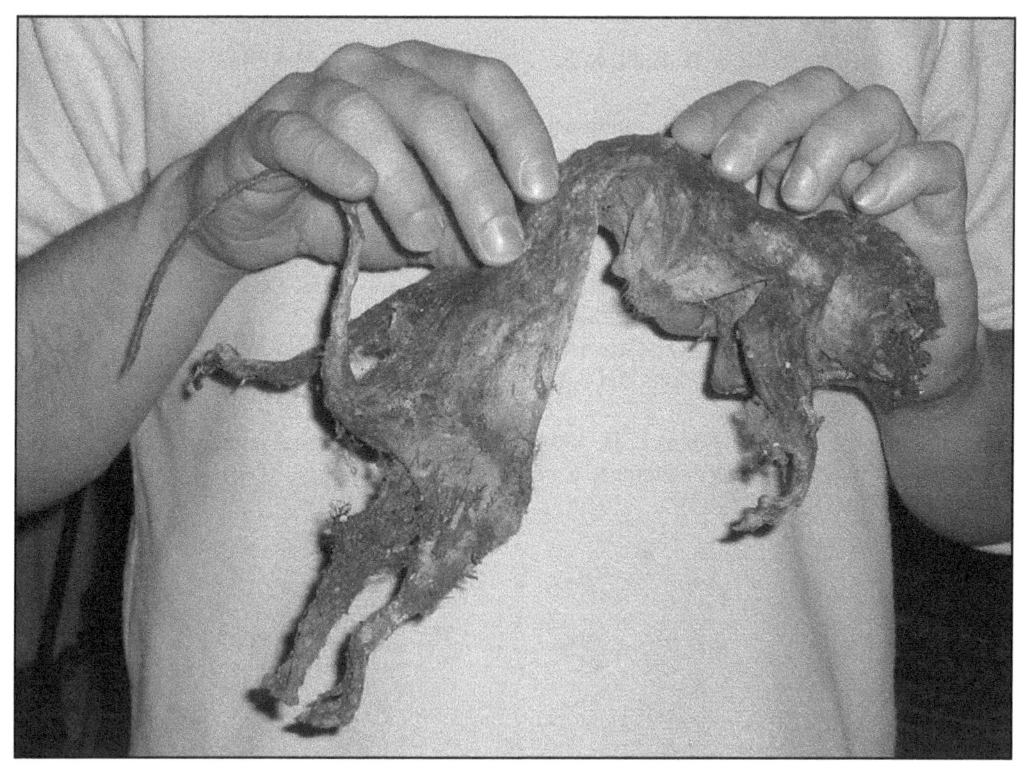

Fig. 80. My wind-dried squirrel corpse.

So, thanks to its tendon locking mechanisms, the dead robin still clings to its perch. Tendon locking mechanisms also mean that bats that die while hibernating also stay fixed to their point of purchase. As Schweitzer *et al.* (2003) wrote: "They may hang all night long, during hibernation or still after they have died" (p. 70). In fact, even when the dead bat's body has rotted and the bones have fallen to the cave floor, the feet can remain secured to the roof. Moving now to cases that don't involve tendon locks, I've heard of sloth carcasses that have been seen lodged up in trees, and Knott (1998) reported a case where an orangutan died slumped on a horizontal log. Frank Buckland owned a mummified marmoset that had been discovered within the hollow branch of a tree (he took it with him when he went to have dinner with Bishop Wilberforce one day), but I'm not sure if the marmoset actually died in the branch or was placed there.

Refs - -

- Gurnell, J. 1987. *The Natural History of Squirrels*. Christopher Helm (London).
- Haffner, M. 1996. A tendon locking mechanisms in two climbing rodents, *Muscardinus avellanarius* and *Micromys minutus* (Mammalia, Rodentia). *Journal of Morphology* 229, 219-227.
- Hill, J. E. & Smith, J. D. 1984. *Bats: a Natural History*. British Museum (Natural History) (London).
- Knott, C. 1998. Orangutans in the wild. *National Geographic* 194 (2), 30-57.
- Quinn, T. H. & Baumel, J. J. 1990. The digital tendon locking mechanism of the avian foot (Aves). *Zoomorphology* 109, 281-293.
- Quinn, T. H. & Baumel, J. J. 1993. Chiropteran tendon locking mechanism. *Journal of Morphology* 216, 197-208.
- Schutt, W. A. 1993. Digital morphology in the Chiroptera: the passive digital lock. *Acta Anatomica* 148, 219-227.
- Schweitzer, A., Frank, O., Ochsner, P. E. & Jacob, H. A. C. 2003. Friction between human finger flexor tendons and pulleys at high loads. *Journal of Biomechanics* 36, 63-71.
- Simmons, N. B. & Quinn, T. H. 1994. Evolution of the digital tendon locking mechanism in bats and dermopterans: a phylogenetic perspective. *Journal of Mammalian Evolution* 2, 231-254.

Chapter 28:
Mystery birds of the Falkland Islands

In Chapter 26 I discussed the little-known case of the Odedi, a cryptic passerine from Bougainville Island in the SW Pacific: it had been heard many times before a specimen was finally procured and the species was officially recognised. If that sort of thing interests you, you might then be pleased to learn that there are actually quite a few little crypto-birds of this sort, mostly unknown to all but specialists.

Unidentified owls on the Mascarenes, the Andaman and Nicobar islands; a black, long-tailed passerine, notable for its startling rattle-like call, from Goodenough Island in the D'Entrecastaeux Archipelago; mystery African gallinules and touracos; a Kenyan long-tailed passerine with reddish undertail-coverts; an all-black Kenyan swift (Ali & Ripley 1969, Williams & Arlott 1980, Beehler 1991, Shuker 1998). There are others that are even less well reported: a brown, thrush-sized duck from St. Paul Island in the Indian Ocean (described by John Barrow in 1793); an unverified serin from Ethiopia, referred to in passing by Clement *et al.* (1993, p. 17); and... the mystery birds of the Falkland Islands.

The Falkland Islands and their native fauna

Recently I've been reading Robin W. Woods' *Guide to Birds of the Falkland Islands*. It's not just a field guide: Woods (an outstanding field ornithologist and collector of data*) included stacks of information on Falkland ecology, topography, and ornithological history. Consisting of 780 islands (only two of which – East Falkland and West Falkland – can be considered large), the Falklands are located about 500 km northeast of continental South America. There are some really strange landscape features: the stone-runs for example, which are accumula-

*And also a psychologist, metereologist and ecologist by the way.

tions of large, angular boulders arranged on the sides of hills and valleys, and there are no native trees (though many trees have of course been planted by human colonists).

Mostly the islands are covered by what's known as oceanic heath: an association of rushes, sedges and mosses that grows on peat layers up to 13 m thick. Tussac grass (or tussock grass, depending on your preference), up to 3.5 m tall, forms dense stands in coastal areas and acts as an important nesting place for petrels, shearwaters, penguins and others. Tussuc is highly sensitive to grazing by large herbivores (introduced cattle all but eradicated it on the larger islands) and, worldwide, tussac species only occur where native herbivorous mammals are rare or absent. Whether the islands were glaciated or not during the Pleistocene remains controversial (McDowall 2005).

As you might expect, there are no lissamphibians or non-avian reptiles native to the islands. There was a native land mammal: the Warrah or Antarctic wolf *Dusicyon australis*, sadly hunted to extinction by 1876. Darwin, who encountered these canids in 1833, famously described how tame and trusting they were. A single bat – a vagrant from Patagonia – has been recorded, and there is also an unverified mention of a small mouse (Day 1981). The avifauna of the islands is pretty good though, with about 36 resident species and 18 additional species recorded as vagrants. The residents are mostly birds of moorland, freshwater environments and shores.

Avian extinctions on the Falklands

It is widely stated in the literature that King penguins *Aptenodytes patagonicus* were made extinct on the Falklands some time around 1870 by the destruction of the last rookery by a shepherd. Apparently he boiled down the birds to use their oil to waterproof a roof, though some authors have doubted the veracity of this story (Simpson 1976). King penguins were thereafter absent from the islands for a while, but by the 1940s they were breeding there and today there are several large colonies.

Less well known is that several birds reported to be present on the islands during the 19th century are no longer there, and hence are assumed to have become locally extinct. Darwin reported or collected Cinereous harriers *Circus cinereus*, Andean tapaculos *Scytalopus magellanicus*, Austral canasteros *Thripophaga anthoides*

Fig. 81. Speckled crake *Coturnicops notata,* as illustrated by John Gould in 1839.

and Yellow-bridled finches *Melanodera xanthrogramma*, all apparently breeding residents. While the harriers were reported by Darwin to be tame residents, by the 1920s they were rare enough to become classified as accidental visitors, and only a handful of sightings have been recorded since the 1960s. The tapaculos haven't been reported since the 1830s and Yellow-bridled finches haven't been reported with confidence from the islands since 1916. The record of the canastero is problematical as Darwin never mentioned this species in *Voyage of the Beagle*, published in 1841, and later authors regarded Darwins's 'Falkland' canasteros as having come from Chile. Maybe he mis-labelled the relevant specimens.

A Speckled crake, perhaps

Moving now to the birds for which specimens weren't retained, Woods (1988) discussed the 1921 capture of a small rail on the bank of a stream near Stanley (the capital, on East Falkland), as reported by Bennett (1926). Was it a Speckled crake *Coturnicops notata* (sometimes called Darwin's rail), as thought by Bennett? If so that's pretty important because the Speckled crake is extremely rare and poorly known, with less than 20 specimens in museum collections (Fig. 81). Bennett's bird died but its skin wasn't kept, and hence its identification was never verified. There don't seem to be any other Speckled crake records from the Falklands, nor records of any other, similar rail.

The 'mystery wrens'

There's more. The Falklands also have a mystery wren, or at least a wren-like passerine. Reported several times since 1910, Wace (1921) listed records from Carcass and West Point Islands as well as other places. Woods (1988) discussed the descriptions of these birds that had been passed to him by K. Bertrand. A light eye-ring and russet plumage were mentioned, and while, overall, the birds seemed similar to juvenile House wrens *Troglodytes aedon* or Grass wrens *Cistothorus platensis*, Woods mentioned photographs taken in 1975 that seemed to show a longer tail than that present in House wrens. Further descriptions were provided by A. Douse following 'mystery wren' observations made in April 1987 at Pebble Island and Port Stephen, and Woods ended his discussion of these unidentified passerines by stating "Further careful observations are needed to identify these birds" (1988, p. 225). Obviously, the most recent source I've consulted on this matter is Woods' volume, and I've been unable to deter-

Fig. 82. Cobb's wren *Troglodytes cobbi.*

mine whether these birds have been identified since Woods published this book. Does anybody know? If you're wondering, the 'mystery wrens' are not the same thing as Cobb's wren *Troglodytes cobbi*, a Falkland Island bird that's only recently become recognised as a distinct species in its own right. Named in 1909, it was regarded as a subspecies of the House wren until Woods (1993) argued that it should warrant specific status, a decision since supported by others. As such it's one of only two bird species endemic to the Falklands – the other is the Falkland steamer duck *Tachyeres brachypterus* (take note McDowall (2005), who listed *Tachyeres brachypterus* as the only endemic Falkland Island bird species).

Given their strong similarity to definite wrens, it's almost certain that 'mystery wrens' really are wrens: that is, members of Troglodytidae. By the way (this is directed at European readers who are less likely to be aware of this than Americans) – troglodytids are essentially an entirely American group, and a mostly South American one at that given that there are only nine species in North America compared to nearly 70 in South America (and it's actually Central America that is the center of their diversity). Only one species has colonized Eurasia, *Troglodytes troglodytes* (we just call it the Wren of course, but if you're American it's the Winter wren). As elucidated by molecular data, its biogeographical history is actually bizarrely complicated (Drovetski *et al.* 2004).

Finally... a possible rayadito

Anyway, to get back to the Falklands birds, some of the other wren-like birds reported from the islands were almost certainly *not* troglodytids. Take the small passerines observed during the 1930s by C. Bertrand on East Sea Lion Island. They were smaller than wrens, 'frail' in appearance, possessed an obvious yellow eye-stripe, and moved rapidly up and down tussac stems. None of the verified Falkland passerines look like this, so Woods (1988) suggested that they might have been Thorn-tailed rayaditos *Aphrastura spinicauda.*

Thorn-tailed rayaditos are quite common in temperate southern South America, and they more or less match Bertrand's description. Perhaps they were vagrants to the Falklands, or (as with the Andean tapaculo and others discussed above) maybe they were natives that have since become extinct. Though it's a woodland bird, it will apparently make do in areas where there is shrubby vegetation, so it's certainly possible that they would be ok on the Falklands.

What are rayaditos? They're furnariids (ovenbirds), but fur-

Fig. 83. Thorn-tailed rayadito *Aphrastura spinicauda.* Photo by Arthur Chapman.

nariids that have evolved to live in temperate woodland. Given the really cool work on furnariid ecomorphological diversity, adaptational shifts, phylogeny and nest diversity that's recently been published (see Fjeldså *et al.* 2005 and Irestedt *et al.* 2006), I'd like to say a lot more about them, but that will have to wait.

Refs - -

- Ali, S. & Ripley, S. D. 1969. *Handbook of the Birds of India and Pakistan, Together With Those of Nepal, Sikkim, Bhutan and Ceylon, Vol. 3*. Oxford University Press (Oxford and Bombay).
- Beehler, B. M. 1991. *A Naturalist in New Guinea*. Texas University Press (Austin, Texas).
- Bennett, A. G. 1926. A list of the birds of the Falkland Islands and dependencies. *Ibis* 2, 306-333.
- Clement, P., Harris, A. & Davis, J. 1999. *Finches & Sparrows*. Christopher Helm (London).
- Day, D. 1989. *The Encyclopedia of Vanished Species*. Universal Books (London).
- Drovetski, S. V., Zink, R. M., Rohwer, S., Fadeev, I. V., Nesterov, E. V., Karagodin, I., Koblik, E. A. & Red'kin, Y. A. 2004. Complex biogeographic history of a Holarctic passerine. *Proceedings of the Royal Society of London B* 271, 545-551.
- Fjeldså, J., Irestedt, M. & Ericson, P. G. P. 2005. Molecular data reveal some major adaptational shifts in the early evolution of the most diverse avian family, the Furnariidae. *Journal of Ornithology* 146, 1-13.
- Irestedt, M., Fjeldså, J. & Ericson, P. G. P. 2006. Evolution of the ovenbird-woodcreeper assemblage (Aves: Furnariidae) – major shifts in nest architecture and adaptive radiation. *Journal of Avian Biology* 37, 260-272.
- McDowall, R. M. 2005. Falkland Island biogeography: converging trajectories in the South Atlantic Ocean. *Journal of Biogeography* 32, 49-62.
- Shuker, K. P. N. 1998. A supplement to Dr Bernard Heuvelmans' checklist of cryptozoological animals. *Fortean Studies* 5, 208-229.
- Simpson, G. G. 1976. *Penguins: Past and Present, Here and There*. Yale University Press (New Haven and London).
- Wace, R. H. 1921. Lista de aves de las isles Falkland. *El Hornero* 2, 194-204.
- Williams, J. G. & Arlott, N. 1980. *A Field Guide to the Birds of East Africa*. Collins (London).
- Woods, R. W. 1988. *Guide to Birds of the Falkland Islands*. Anthony Nelson (Oswestry).
- Woods, R. W. 1993. Cobb's Wren *Troglodytes (aedon) cobbi* of the Falkland Islands. *Bulletin of the British Ornithologists' Club* 113, 195-207.

Chapter 29:
In quest of anguids

The anthropogenic extinction of bears, moose, aurochs, beavers, wolves, lynx, bustards and everything else bigger than a house brick means that Britain has no 'charismatic megafauna' (domestics and recent introductions notwithstanding). But given that all tetrapods are inherently fascinating, our fauna still has a lot going for it, and we still have a lot of neat beasts. It's near the top of the list – it's the Slow-worm *Anguis fragilis*, a limb-less anguid also known as the Blind-worm or Ovet (apparently). It occurs throughout Europe (except for southern Portugal and Spain, Corsica and Sardinia, Ireland*, and much of Scandinavia) and is also found in Asia as far east as the Caucasus, and in parts of Algeria and Tunisia.

Fig. 84. Female Slow-worm *Anguis fragilis* on my hand. Note the scars (on the slow-worm).

---

* Although there are some indications that the species might actually be present on Ireland. See McCarthy (1977).

In my experience lay people find it utterly inconceivable that it's not a snake, and most think that it's a 'grass snake' (the true Grass snake *Natrix natrix* is an amphibious colubrid that's normally between 60-100 cm long). Slow-worms are nocturnal and semi-fossorial and mostly occur in well-vegetated places with thick ground cover and loose soils. They hide under large stones, pieces of wood, sheet metal and other debris. Of course places that fulfill these criteria are exactly those places that are nowadays covered by concrete or turned into housing estates, and today there are less gardens and allotments that are suitable for slow-worms than there used to be.

They also suffer heavily from predation by domestic cats: over a five month period of a single year, Britain's 9 million cats are estimated to bring home about 5 million non-avian reptiles and lissamphibians, and of these about 12% are slow-worms (Woods *et al.* 2003). That amounts to about 600,000 slow-worms killed over five months, every year, which might be a non-sustainable number if we actually knew what the total population is (and we don't). Certainly my parent's cats reportedly kill or mortally wound enough slow-worms to create the impression that all the large juveniles and adults in the area are being extirpated, and slow-worms are certainly less common here in Hampshire now than they were when I was a boy.

Despite this, however, I'm pleased to say that even today it's not difficult to find them in suitable places if you know where to look, and lately Will and I have discovered them whenever we've search in the right kind of habitat. On a well-vegetated piece of waste-ground about 200

Fig. 85. Three slow-worms discovered under a bit of old carpet. Two males are on the right; a female is on the left.

m from my front garden there is a piece of carpet where there's an ant's nest, and it's under this piece of carpet that we've been finding slow-worms. Last week I was lucky enough to discover three, all hiding there together, and they're the individuals featured in the photos here. Slow-worms are viviparous, giving birth to 5-12 babies (though exceptionally as many as 26) during August or September. The juveniles, normally around 10 cm long, have a gold dorsum with a black central dorsal stripe and black sides. In adults, it's easy to distinguish males from females. Females tend to have a proportionally smaller head and more distinct neck, and also a more demarcated tail, but more obviously they have a dark stripe down the centre of the back and dark sides. The dorsal stripe is usually straight but in some individuals it zig-zags.

Individuals of both sexes are often scarred: females, because they get gripped around the head and neck by males during mating, and males, because they bite one another vigorously when fighting. As you can seen from the photos, we found one female and two males (Fig. 85). Note the scars on the female (Fig. 84). Their size, shape and distribution matches what you'd expect had she been bitten by males. Note also the lost and partially regrown tail tip. Slow-worms are like many other squamates in being able to drop the tail voluntarily by contracting muscles (this is known as caudal autotomy), but they do this more than is usual, with most individuals having autotomised their tails several times. This explains the specific name by the way.

Why slow-worms exhibit prominent dimorphism I don't know, but it's been suggested that the dorsal stripe of the females might lend them some superficial similarity to the Adder *Vipera berus*, as this venomous species also has a zig-zag dark stripe along its dorsal midline. This is utterly unconvincing, however, given that the zig-zagging is indistinct, and that slow-worms and adders are otherwise so utterly different in shape and size. Furthermore, this theory is horribly anglocentric as it's not as if the adder is the only venomous snake that inhabits the same range as the slow-worm (in Britain, adders are the only venomous native snake). In fact, other European snakes with zig-zagging dark dorsal stripes (e.g., the Viperine snake *Natrix maura*) are no more dangerous to potential predators than the slow-worm is, so mimicking them wouldn't really be such a good thing. Oh yeah, among the most important predators of slow-worms are other squamates (most notably the Smooth snake *Coronella austriaca*), and they identify their slow-worm prey by olfaction, not by visual clues.

Large, old male slow-worms are sometimes decorated with light blue spots, and at least a few females possess these spots too. The spots vary in intensity, distribution and number, and appear to occur randomly among populations. In Britain it used to be thought that blue-spotted animals represented a distinct subspecies that was dubbed *Anguis fragilis colchica*, but in fact there is no such taxonomic entity. To confuse things however, there is a supposed anguid taxon endemic to eastern Europe that was also named *A. f. colchica* (that was the original name – coined by Nordman in 1840 – but the alternative spelling *A. f. colchicus* has also been used at times), and it also is characterized, in part, by its blue spots (which are larger and more prominent than those of other slow-worms). While most authors continue to regard *A. f. colchica* as a valid taxon, Šandera *et al.* (2004) found supposed *A. f. colchica* individuals living sympatrically with *A. f. fragilis* individuals on Corfu, and they hence doubted the idea that *A. f. colchica* really warrants taxonomic separation. Southern Greece is home to a distinct slow-worm population that is best known in the literature as *A. f. peloponnesiacus*, after the Peloponnese Islands where it's found (it also occurs on the southern Ionian Islands). This name was coined in 1937 and it turns out that an older name, *A. f. cephallonica*, was created in 1894 for the same taxon. Distinguishable from other slow-worms by its high scale count at mid-body (Arnold *et al.* 1992), *cephallonica* is now deemed distinct enough from other slow-worms to warrant specific status (as *Anguis cephallonica*: the spelling *A. cephallonicus* is also sometimes used), so nowadays there are actually two slow-worm species.

On the subject of variation and taxonomy, I must say that the English slow-worms I'm familiar with often look quite different from the European and Asian individuals I've seen in photos. Interestingly, non-British slow-worms often look superficially more like glass lizards to me, being proportionally longer, more robust, and with larger, more sharply demarcated cephalic scales. I'm sure that this is an accident resulting from my familiarity with a local group of populations, however, rather than anything important. Having said that, there is a paper titled something like 'Unusual sizes of slow-worms and adders in Britain' which sounds relevant to this, but I cannot find the full reference, let alone the paper, dammit. And on the subject of slow-worms and their close kin the glass lizards, I was going to discuss anguid phylogeny, the evolution of limblessness and the tricky issue of glass lizard paraphyly, but that'll have to wait. Again, if you can't wait until then, then by all means check out the literature I had in mind: Walls (1991), Macey *et al.* (1999), Sullivan *et al.* (1999), Wiens & Slingluff (2001) and Sanger & Gibson-Brown (2004).

I've never known a slow-worm to bite – Knight (1965) wrote that a slow-worm "seldom bites at the human hand that grasps it" (p. 23) – but one that I picked up once opened its mouth and let out a comparatively loud, throaty huff noise. I didn't know they did this, so I asked around, and it seems no one else knows they do this either. I did hear of a case where a pet anguid (I forget of which species) let out a loud shriek, which if accurately reported would have been a genuine vocalisation rather than just a noisy exhalation. As you'll know if you've checked the arcane literature on squamates, which snakes and lizards are really capable of true vocalisation is controversial, with anecdotal records suggesting that the ability to vocalize might be more widely distributed than generally thought. Officially it's only geckos and some lacertids that vocalize, though pygopodids, anoles and teiids have been reported to make distress calls. This is an essentially random distribution within Squamata, so it's certainly conceivable that groups like anguids might have evolved this ability too.

Fig. 86. Slow-worm close-up. Photo by Neil Phillips.

If a slow-worm were to bite, I imagine it would be pretty painful. They prey mostly on small slugs and earthworms, but they also eat small snails, and to crush snail shells you need powerful jaws. Anguids in general have nasty bites: Sprackland (1991) reported that alligator lizards would be able to break human skin.

An average slow-worm is less than 30 cm long. But how big do they get? Well, actually, I don't want to answer this as I'm saving it for another time. It's a story that involves a strange little island in the Bristol Channel and my friend Stig Walsh's back garden.

Refs - -

- Arnold, E. N., Burton, J. A. & Ovenden, D. W. 1992. *Reptiles and Amphibians of Britain and Europe*. Collins (London).
- Knight, M. 1965. *Reptiles in Britain.* Brockhampton Press (Leicester).
- Macey, J. R., Schulte, J. A., Larson, A., Tuniyev, B. S., Orlov, N. & Papenfuss, T. J. 1999. Molecular phylogenetics, tRNA evolution and historical biogeography in anguid lizards and related taxonomic families. *Molecular Phylogenetics and Evolution* 12, 250-272.
- McCarthy, T. K. 1977. The slow-worm, *Anguis fragilis* L.; a reptile new to the Irish fauna. *The Irish Naturalists' Journal* 19, 49.
- Šandera, M., Medikus, S. & Šanderová, H. 2004. Poznámky k herpetofauné Korfu a ostrava Paxos. *Herpetologicke Informace* 1/2004, 24-30.
- Sanger, T. J. & Gibson-Brown, J. 2004. The developmental bases of limb reduction and body elongation in squamates. *Evolution* 58, 2103-2106.
- Sprackland, R. G. 1991. Alligator lizards. *Tropical Fish Hobbyist* 39, 136-140.
- Sullivan, R. M., Keller, T. & Habersetzer, J. 1999. Middle Eocene (Geiseltalian) anguid lizards from Geiseltal and Messel, Germany. 1. *Ophisauriscus quadrupes* Kuhn 1940. *Courier Forschungsinstitut Senckenberg* 216, 97-129.
- Walls, J. G. 1991. The name game: alligator lizards and glass "snakes". *Tropical Fish Hobbyist* 39, 142-145.
- Wiens, J. J. & Slingluff, J. L. 2001. How lizards turn into snakes: a phylogenetic analysis of body-form evolution in anguid lizards. *Evolution* 55, 2303-2318.
- Woods, M., McDonald, R. A. & Harris, S. 2003. Predation of wildlife by domestic cats *Felis catus* in Great Britain. *Mammal Review* 33, 174-188.

Chapter 30:
Pterosaur wings: broad-chord, narrow-chord, both, in-between, or... all of the above?

What did pterosaurs look like when they were alive? Did they have relatively broad wing membranes [patagia from hereon] that – like those of most bats – stretched as far as their ankles, or did they have narrow patagia that attached to their hips, or perhaps to the tops of their thighs? Until the 1980s, pterosaurs were pretty much universally depicted as possessing broad-chord wings that extended to their ankles. There were exceptions: K. A. von Zittel (1882) imagined pterosaurs as possessing narrow, swallow-like wings that did not attach further distally than the knee, and Harry Seeley (1901) opined that the patagia may not have incorporated the hindlimbs at all. But, mostly, pterosaurs were regarded as bat-winged, an image that Kevin Padian (1987) argued to be primarily typological (viz, pterosaurs were initially imagined as bat-like, therefore they must have had bat-like patagia). It's well known that, following his studies of *Dimorphodon* and other pterosaurs, Padian championed the idea that pterosaurs were agile cursorial bipeds, and that their patagia were narrow and did not incorporate the hindlimbs (Padian 1983). This idea was popular for a while but, judging from the way pterosaurs are depicted today, it doesn't seem that popular now (though this isn't to say that there aren't scientists and artists who *are* reconstructing pterosaurs in this manner).

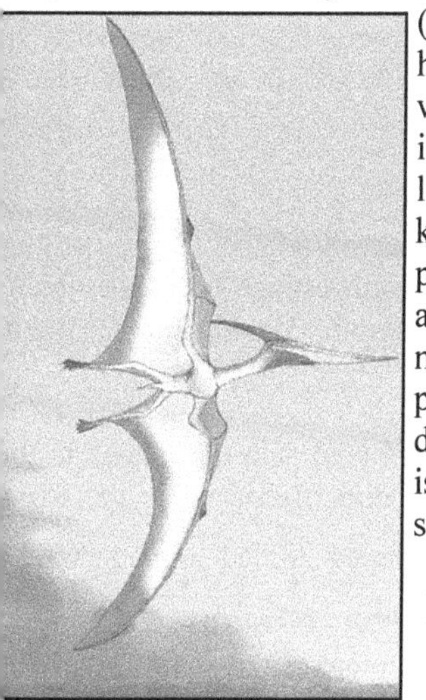

Fig. 87. *Pteranodon*, shown with broad patagia that attach to the ankles. Image by Mark Witton.

If anything seems popular, it's actually a sort of hybrid morphology: while the patagia are ordinarily depicted as incorporating the hindlimb, they aren't shown as extending as far as the ankle, but to the knee or shin. This goes for all of John Sibbick's pterosaurs (in Wellnhofer 1991 and elsewhere) as well as for many other popular life restorations. I'll admit that my personal aesthetic sense of what is, and what is not, cool leads me to intuitively prefer the narrow-chord model, and as a teenage enthusiast self-teaching myself from the writings of Bob Bakker and Greg Paul I personally took a liking to narrow-chord wings. But being a scientist is all about going where the evidence leads, and not on what you might intuitively prefer. Where does the evidence lead: did pterosaurs have broad-chord or narrow-chord wings? Or both?

Fig. 88. The azhdarchid *Zhejiangopterus* shown with narrow patagia. Image by John Conway.

Having considered this matter in some depth (Naish & Martill 2003), and having looked at some of the pertinent specimens, I conclude that broad-chord patagia are best supported, and in fact are *well* supported. So, yes, the data best supports inclusion of the hindlimbs in the patagia, and extension of the patagia as far distally as the ankle. Shock horror: that's a pretty controversial point of view in the world of pterosaur research. But quite a few specimens show this to have been the case. Let's look at some of them.

-- The famous 'dark wing' *Rhamphorhynchus* specimen (Fig. 89) preserves intact, in-situ patagia, and the left brachiopatagium "curves caudomedially and attaches to the ankle of the left hindlimb" (Frey *et al.* 2003, p. 243). This is clearly visible in good photos of the specimen and there is no question that the patagia are genuine, intact and preserved in-situ. I have access to an excellent cast of this three-dimensional specimen and am happy with Frey *et al.*'s interpretation.

-- The *Eudimorphodon* specimen MCSNB 8950 has patagia preserved adjacent to the ankles of both legs. Fibres preserved within the membranes appear to confirm brachiopatagia attaching as far distally as the ankle, and the presence of a cruropatagium (Bakhurina & Unwin 2003).

-- The holotype of little *Sordes pilosus* indisputably preserves brachiopatagia that attach to the ankles (Unwin & Bakhurina 1994). Claims that the patagial margins preserved in the fossil actually represent cracks are utterly unconvincing.

-- The Crato Formation azhdarchoid specimen SMNK Pal 3830 shows a brachiopatagium extending distally to contact the ankle (Frey *et al.* 2003). While this pterosaur is incomplete, the patagial edge is continuous, smoothly concave, and grades neatly into the side of the tarsus. I've examined the specimen up-close in person, and was personally happy that the specimen really does preserve a partial brachiopatagium and really does preserve a brachiopatagium that attached at the ankle. In fact it was the key specimen that convinced me of the reality of broad-chord patagia [Incidentally, the phylogenetic affinities of this specimen are uncertain. Frey *et al.* (2003) regarded as it a possible azhdarchid, but it possesses certain features suggesting that it might be more closely related to *Tupuxuara*].

While the interpretations given above have been challenged for some of these specimens (Peters 1995, 2001), I note the following: ALL the pterosaur specimens for which we have the patagia preserved appear to show broad-chord patagia and, while in some of these cases the broad-chord interpretation is ambiguous or arguable, there are NO specimens that unambiguously preserve narrow-chord patagia. Claims that narrow-chord membranes, unattached to the distal hindlimb or even unattached to the hindlimb altogether, are preserved in some specimens (e.g., the anurognathid *Jeholopterus* and the Zittel wing *Rhamphorhynchus*) are either based on data that is even more ambiguous than that discussed above, or are erroneous (e.g., the Zittel wing can't be used to demonstrate lack of attachment to the hindlimb, as the caudolateral part of the membrane is missing and we don't know how extensive the membrane was when complete).

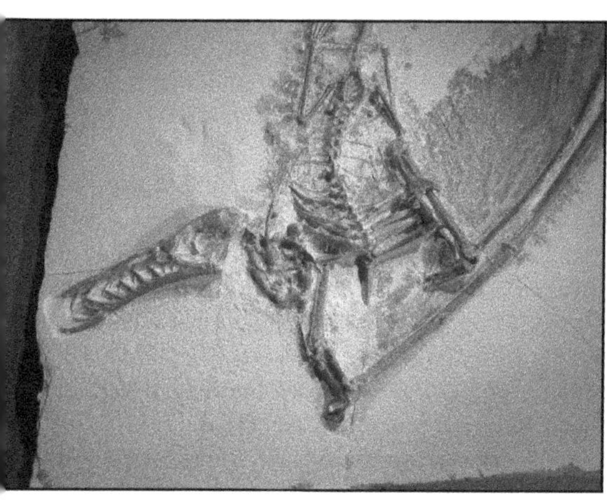

Fig. 89. The 'dark wing' *Rhamphorhynchus* specimen. Photograph by kind courtesy of Dave Hone.

Data from pterosaur hindlimb proportions provides support for the idea that the hindlimbs were incorporated into the patagia. By plotting the lengths of femora, tibiae and metatarsi onto ternary diagrams, Daniel Elvidge and David Unwin found that pterosaurs occupied a tight, compact group of data points within morphospace, and a 'data cloud' similar in size to that occupied by bats. The cloud occupied by birds was more than twice the size of the pterosaur or bat clouds (Elvidge & Unwin 2001). This data indicates that pterosaurs were constrained in hindlimb proportions in the same manner that bats are. I find the most plausible explanation for this to be the linking of the fore- and hindlimbs by patagia: because pterosaur hindlimbs were always a part of the wing apparatus, pterosaurs did not evolve the diverse hindlimb morphology that birds did.

Significantly (from the point of view of the discussion here), those pterosaur specimens preserved with patagia were distributed randomly within the pterosaur cloud. This indicates that broad-chord patagia are both widely distributed within pterosaurs, and the norm for the group. Dyke *et al.* (2006) have recently argued that broad-chord patagia may not have applied to all pterosaurs, but they seemed unaware that there is evidence for this morphology outside of *Sordes*.

I still think it's at least possible that some pterosaurs had reduced patagia however. Morphological evidence suggests that dsungaripterids were quite terrestrial in habits (Fastnacht 2005), and it's tempting to speculate that they were better suited for walking around on the ground than were others, and hence with less extensive patagia. But, hey, this is part of the reason why pterosaurs are so interesting: they're unique, with no close extant analogues, and this is why they're so controversial.

The evidence we have shows that broad-chord patagia were widespread among pterosaur clades. Maybe there were narrow-chord forms (with my money being on dsungaripterids), but we have yet to find soft-tissue evidence demonstrating their presence.

Refs - -

- Bakhurina, N. N. & Unwin, D. M. 2003. Reconstructing the flight apparatus of *Eudimorphodon. Rivista del Museo Civico di Scienze Naturali "Enrico Caffi"* 22, 5-8.
- Dyke, G. J., Nudds, R. L. & Rayner, J. M. V. 2006. Limb disparity and wing shape in pterosaurs. *Journal of Evolutionary Biology* doi:10.1111/j.1420-9101.2006.01096.x
- Elvidge, D. J. & Unwin, D. M. 2001. A morphometric analysis of the hind-limbs of pterosaurs. In *Two Hundred Years of Pterosaurs, A Symposium on the Anatomy, Evolution, Palaeobiology and Environments of Mesozoic Flying Reptiles. Strata Série* 1 11, 36.
- Fastnacht, M. 2005. The first dsungaripterid pterosaur from the Kimmeridgian of Germany and the biomechanics of pterosaur long bones. *Acta Palaeontologica Polonica* 50, 273-288.
- Frey, E., Tischlinger, H., Buchy, M.-C. & Martill, D. M. 2003. New specimens of Pterosauria (Reptilia) with soft parts with implications for pterosaurian anatomy and locomotion. In Buffetaut, E. & Mazin, J.-M. (eds) *Evolution and Palaeobiology of Pterosaurs. Geological Society Special Publication 217*. The Geological Society of London, pp. 233-266.
- Naish, D. & Martill, D. M. 2003. Pterosaurs – a successful invasion of prehistoric skies. *Biologist* 50, 213-216.
- Padian, K. 1983. A functional analysis of flying and walking in pterosaurs. *Paleobiology* 9, 218-239.
- Padian, K. 1987. The case of the bat-winged pterosaur: typological taxonomy and the influence of pictorial representation on scientific perception. In Czerkas, S. J. & Olson, E. C. (eds) *Dinosaurs Past and Present Vol. II*. Natural History Museum of Los Angeles County/University of Washington Press (Seattle and London), pp. 64-81.
- Peters, D. 1995. Wing shape in pterosaurs. *Nature* 374, 315-316.
- Peters, D. 2002. A new model for the evolution of the pterosaur wing – with a twist. *Historical Biology* 15, 277-301.
- Seeley, H. G. 1901. *Dragons of the Air: An Account of Extinct Flying Reptiles.*

Methuen (London).
- Unwin, D. M. & Bakhurina, N. N. 1994. *Sordes pilosus* and the nature of the pterosaur flight apparatus. *Nature* 371, 62-64.
- Wellnhofer, P. 1991. *The Illustrated Encyclopedia of Pterosaurs*. Salamander Books (London).
- Zittel, K. A. von 1882. Über Flugsaurier aus dem lithographischen Schiefer Bayerns. *Paläontographica* 29, 47-80.

Chapter 31:
The most freaky of all mammals: rabbits

Many animals which we take for granted are, when you think about them, actually very odd. And for a long time I've been thinking that this is oh so true of one of the mammals I see the most, the rabbit *Oryctolagus cuniculus*. Actually, I don't have that species specifically in mind, but in fact all lagomorphs. Before I start on the generalizations, I'll take this opportunity to point out (for those who might not know) that – while we have millions of bunnies here in the UK – they're not native. The rabbit is in fact an animal of the Mediterranean region, and it's supposed to have been introduced by the Normans after the conquest of 1066. However, there is apparently no mention of rabbits in the Domesday Book (written in 1086), and they don't get a mention in the literature until 1176 (and even then only in a report about the Scilly Isles). It's on the basis of this that some workers think it more likely that rabbits were actually introduced by the Crusaders in the 12th century (McBride 1988). We do have two native lagomorphs by the way, the Brown hare *Lepus europaeus** and the Blue hare *L. timidus**. Within Lagomorpha, rabbits and hares make up the clade Leporidae, and the less well-known pikas (aka ochotonids) form their sister-taxon.

Fig. 90. Rabbit! Specifically, European rabbit *Oryctolagus cuniculus*. Photo by Neil Phillips

* The correct species name for the Brown hare is controversial and the reality/monophyly of the Blue hare has recently been contested. I don't want to cover these issues here: if you're interested see Waltari & Cook (2005) and Ben Slimen *et al*. (2006).

Musings on *Watership Down*

It's not in line with the rest of this article, but I'll never forgive myself if I miss this opportunity to talk about *Watership Down.* I must confess to never having read Richard Adams's 1972 book, but I really like the film and I like it more the more I see it. While not exactly zoologically accurate (the rabbits have religion, mythology, language, human emotions, team up with a friendly Black-headed gull *Chroicocephalus ridibundus*, and learn how to use boats, among other things), some of it is not a million miles away from what we really know about rabbit society.

Two things make the film particularly memorable. Firstly, it begins with the rabbit myth of creation. In the beginning Frith, the lord of creation, made all animals alike, and they ate grass together. But El-ahrairah, the first rabbit, produced so many children that Frith became angry: control your people, or I will do something about it. El-ahrairah did nothing about it, so Frith did. He gave each of the animals a gift, and they were no longer the friends of El-ahrairah's children – they wanted to catch them and kill them. The weasel. The stoat. The fox. The hawk. The owl. But Frith also gave El-ahrairah a gift – a bright white tail that flashed as a warning, long legs to run fast, and big ears to hear his enemies. The sequence ends as Frith tells El-ahrairah "All the world shall be your enemy, prince with a thousand enemies, and whenever they catch you, they will kill you. But first they must catch you, digger, listener, runner, prince with a swift warning. Be cunning and full of tricks and your people shall never be destroyed". It's a good way to start a film. The second thing that makes the film memorable is how dark and disturbing it is in places. Some of the main rabbit characters are killed by predators, caught in snares, or buried alive. Death, overall, is an important part of the story, as it is in the real lives of rabbits I suppose.

Why rabbits are just wrong

Moving on… why regard rabbits as 'the most freaky of all mammals'? To begin with, just look at how weird they are. They're familiar to us, but their anatomy is actually highly odd. Example? Their teeth are strange, with cusps and folds that have proved almost impossible to homologise with those of other placental mammals. The sides of their snout bones are decorated with a bizarre lattice-work of filigree bone texture (Fig. 91). Their incisive foramina (openings on the bony palate) are uniquely elongate. A thin splint from their frontal bones projects down and forward, finger-like, among the snout bones. Their hindlimbs are proportionally elongate relative to their forelimbs (odd for a quadupedal mammal, when you think about it). Their ankle bones are uniquely strange, with the calcaneum housing a canal that runs diagonally

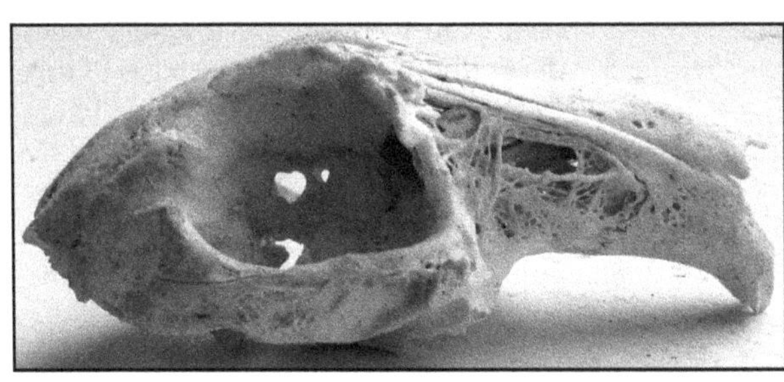

Fig. 91. A rabbit skull, showing the weird, filigree bone texture on the side of the snout.

through the bone (Bleefeld & Bock 2002). No other mammal has anything remotely like this. The undersides of their feet are completely covered by thick fur – that's odd, and unique. And don't get me started on their genitals (read on).

Many aspects of their physiology and behaviour are also odd compared to what we're more familiar with. Get this: when baby rabbits suckle, the milk is ejected in one big squirt, only triggered after the mother has been sufficiently stimulated by the paddling action of the babies' paws. Male rabbits also squirt, but this time the liquid is urine, and it gets squirted over potential mates. As is reasonably well known, lagomorphs practice refection – that is, they have to 'rescue' nutrients from their digested food by ingesting their own caecal pellets (they therefore only produce dry droppings once the food has been through the system twice). And lagomorphs are also odd in practicing so-called absentee care, with mother rabbits spending just 0.1% of their time with their young.

What, if anything, is a rabbit? (homage to Wood)

Working on the assumption that organisms should be regarded as freaky when we can't even work out what they are, rabbits excel. Albert Wood (1957) explored this area when he wrote 'What, if anything, is a rabbit?'. Check out the first paragraph of his paper: "The title of this paper is slightly modified from that of an article I encountered some years ago, which appeared to be approaching the problem of the relationships of the Lagomorpha, or rabbits and their relatives, from the most basic point of view. This paper, entitled "Gibt es Leporiden?", seemed to be questioning the very existence of such animals. Investigation showed, however, that the question involved was not whether members of the family Leporidae existed, but whether rabbit-hare hybrids did. Since then, I have met no one who questions the existence of rabbits and hares, and I have been reluctantly forced to accept them" (Wood 1957, p. 417).

Fig. 92. Another rabbit, this time an Eastern cottontail *Sylvilagus floridanus*.

Originally, rabbits were included in Rodentia, and they weren't formally separated from them until Gidley (1912) did so. What makes this decision particularly interesting is that Gidley suggested that lagomorphs had no close relationship to rodents at all, but shared some intriguing similarities with artiodactyls. While a few authors commented on this idea after Gidley, the evidence for it isn't great. Mostly it comes down to a superficial similarity between certain Cenozoic artiodactyls (like cainotheres) and lagomorphs, and the transverse chewing style and artiodactyl-like ankle structure of lagomorphs.

It's also been noted that lagomorphs possess similarities with the pantodonts and dinoceratans of the Palaeogene. If you know what the members of these groups looked like, you'll understand why positing an affinity between them and lagomorphs is so radical. I'll cover it some other time.

Lagomorph ancestry has also been sought among the various hoofed mammals collectively termed 'condylarths', and in particular they've been tied to periptychids like *Ectoconus*. For reasons of time and space I don't want to expand on this point either, but I may do so later.

Fig. 93. One of the weirdest of the weird: the North American Pygmy rabbit *Brachylagus idahoensis*.

Bunnies: Mesozoic relicts, or para-marsupials?

In keeping with the idea that lagomorphs have no close living relatives, it has been proposed at times that they might have descended from groups that were otherwise entirely restricted to the Mesozoic. Based on tooth cusp morphology, Gidley (1906) suggested that lagomorphs descended from triconodontids. McKenna (1982, 1994) argued that lagomorphs are part of a larger placental clade [termed Anagalida in McKenna & Bell (1997)] that includes as its most basal members the Cretaceous zalambdalestids, although the evidence for this has more recently been assessed and rejected.

Most zoologists would be surprised to hear that, in a few features, lagomorphs resemble marsupials more than they do placentals, and it's on the basis of these features that some workers have actually suggested that lagomorphs might be close kin of marsupials (albeit not necessarily members of Marsupialia or even Metatheria). Gregory (1910) drew attention to the arterial foramen present in the last cervical vertebra, supposedly uniquely shared by lagomorphs and marsupials (but actually occurring more widely among placentals); Hartman (1925) showed that egg development in the lagomorph fallopian tube was uniquely marsupial-like; and Petrides (1950) pointed out that lagomorphs are unique among placentals in possessing a pre-penile scrotum, a character also otherwise limited to marsupials. That's right, a pre-penile scrotum. A scrotum that is further away from the anus than the penis is. I've actually, err, manipulated a few rabbits to observe this remarkable configuration, thus far without success, but then the individuals in question were neutered. Hmm. Anyway: so, are rabbits actually some long-lost freakish sister-group to metatherians?

The primate hypothesis, and the resurrection of Glires

Protein sequences led Graur *et al.* (1996) to argue that lagomorphs were closest to primates, and these authors further argued that morphological characters used to unite lagomorphs with rodents and other groups were not really indicative of affinity. This hypothesis was hailed at the time as the most likely answer to Woods' 'What, if anything, is a rabbit?', but it suffers from that widespread problem of assuming that one body of evidence must somehow outweigh, or be superior to, all the other data.

The most recent assessment of the morphological and fossil data indicates that lagomorphs are, after all, most closely related to rodents, with the two forming the larger clade Glires. This is supported by the detailed morphology of Palaeocene proto-lagomorphs like *Gomphos*, and by a big data set with good character support across nodes (Asher *et al.* 2005). According to recently published phylogenetic definitions, the term Lagomorpha is best restricted to the pika-rabbit clade (viz, the crown-clade) and the old name Duplicidentata is applied to the stem-group that includes Lagomorpha. If you're interested, both morphological and molecular data supports the inclusion of Glires within the more inclusive clade Euarchontaglires, and herein there are the primates, dermopterans and tree shrews.

So after all that, rabbits really do seem to be part of a clade that is closest to rodents. Sadly, they aren't para-marsupials, close kin of *Uintatherium*, or relict survivors from a long-lost Cretaceous radiation. But I still think they're freaky.

Refs - -

- Asher, R. J., Meng, J., Wible, R. R., McKenna, M. C., Rougier, G. W., Dashzeveg, D. & Novacek, M. J. 2005. Stem Lagomorpha and the antiquity of Glires. *Science* 307, 1091-1094.
- Ben Slimen, H., Suchentrunk, F., Memmi, A., Sert, H., Kryger, U., Alves, P. C. & Ben Ammar Elgaaied, A. 2006. Evolutionary relationships among hares from north Africa (*Lepus* sp. or *Lepus* spp.), cape hares (*L. capensis*) from South Africa, and brown hares (*L. europaeus*), as inferred from mtDNA PCR-RFLP and allozyme data. *Journal of Zoological Systematics* 44, 88-99.
- Bleefeld, A. R. & Bock, W. J. 2002. Unique morphology of lagomorph calcaneus. *Acta Palaeontologica Polonica* 47, 181-183.
- Gidley, J. W. 1906. Evidence bearing on tooth-cusp development. *Proceedings of the Washington Academy of Science* 8, 91-110.
- Gidley, J. W. 1912. The lagomorphs an independent order. *Science* 36, 285-286.
- Graur, D., Duret, L. & Guoy, M. 1996. Phylogenetic position of the order Lagomorpha (rabbits, hares and allies). *Nature* 379, 333-335.
- Gregory, W. K. 1910. The orders of mammals. *Bulletin of the American Museum of Natural History* 27, 1-524.
- Hartman, C. G. 1925. On some characters of taxonomic value appertaining to the egg and ovary of rabbits. *Journal of Mammalogy* 6, 114-121.
- McBride, A. 1988. *Rabbits & Hares*. Whittet Books (London).
- McKenna, M. C. 1982. Lagomorph interrelationships. *Geobios, mémoire spécial* 6, 213-223.
- McKenna, M. C. 1994. Early relatives of Flopsy, Mopsy, and Cottontail. *Natural History* 103 (4), 56-58.
- McKenna, M. C. & Bell, S. K. 1997. *Classification of Mammals: Above the Species Level.* Columbia University Press (New York).
- Petrides, G. A. 1950. A fundamental sex difference between lagomorphs and other pla-

cental mammals. *Evolution* 4, 99.

- Waltari, E. & Cook, J. A. 2005. Hares on ice: phylogeography and historical demographics of *Lepus arcticus*, *L. othus*, and *L. timidus* (Mammalia: Lagomorpha). *Molecular Ecology* 14, 3005-3016.
- Wood, A. E. 1957. What, if anything, is a rabbit? *Evolution* 11, 417-425.

Chapter 32:
Dicynodonts that didn't die: late-surviving non-mammalian synapsids I

Rightly or wrongly, I've invested an unreasonable amount of time in arguing against what's been termed the 'prehistoric survivor paradigm': the notion, endorsed and propounded by some cryptozoologists, that numerous tetrapod groups known only from the fossil record might have survived to the present, yet without leaving an intervening fossil record. But the history of zoological discovery shows us that there's nothing intrinsically wrong with the idea of either long gaps in the fossil record, or of the discovery of Lazarus taxa (that is, organisms that represent late survivors of supposedly long-extinct groups).

We now know that quite a few fossil tetrapod groups really did hang on for much longer than was conventionally thought: in these cases, both long gaps in the fossil record, and the unexpected appearance of Lazarus taxa, occur. Some such cases have been big news in the palaeontological world, but mostly these instances have been largely ignored or unreported given that the animals concerned were obscure, deemed mundane, or of interest only to specialists. I personally think all the relevant cases are really interesting, and in future writing I aim to review them. Here we're going to look at some of the more obscure and unreported of these cases – namely, the late survival of non-mammalian synapsids. If you don't know what these are, the following paragraph is for you. If you do, skip ahead.

Mammals are the only surviving members of a far larger and more diverse tetrapod clade: Synapsida. Mammals evolved in the Triassic*, but they are but one of numerous synapsid clades, few of which made it past the mass extinctions at the end of the Triassic, and most of which evolved and diversified in the Carboniferous and Permian. Basal synapsids (things like caseids, ophiacodontids and sphenacodontids) looked superficially reptile-like, and for this reason non-mammalian synapsids have traditionally been dubbed 'mammal-like reptiles', and even classified within Reptilia. This obscures their affinity with mammals, and it's nowadays agreed that Synapsida and Reptilia are different clades: basal synapsids are better regarded as 'stem mammals', and they aren't reptiles. While some non-mammalian synapsids were bizarrely unique and quite different from mammals, during the Permian and Triassic a number of very mammal-like non-mammalian synapsids evolved. Some of these animals evolved small body size, and probably endothermy and body fur. So were you to travel back to the Triassic and catch an assortment of small synapsids, you might be hard pressed to work out which were the mammals.

Fig. 94. The Triassic dicynodont *Placerias*. Image by Dmitry Bogdanov.

In terms of taxonomic diversity, one of the most successful groups of non-mammalian synapsids were the dicynodonts, a group that first evolved in the Late Permian and then thrived in the Early and Middle Triassic. Dicynodonts were tubby-bodied, relatively short-legged synapsids (Fig. 94), amusingly described by Mike Benton as possessing an 'unsatisfactory tail'. Basal forms had multiple teeth, but mostly they had a strongly reduced dentition, with only large tusk-like upper canines projecting from their short, beaked jaws. Dicynodonts were probably predominantly herbivorous, but at least some may have chewed at carcasses and eaten small animals.

Sad to say, during the Late Triassic, dicynodonts dwindled in diversity until by the Norian (the penultimate stage of the Late Triassic) they were down to just three genera, and all of these were close relatives within the clade Kannemeyeriiformes (King 1990, Maisch 2001). I always liked Richard Cowen's suggestion that these last forms were ecologically peripheralised, endangered species that hung on to existence in remote ecosystems where life was harsh. In fact Cowen compared them to Giant pandas *Ailuropoda melanoleuca* and Mountain gorillas *Gorilla beringei* if I remember correctly (Martill & Naish 2000 also covered this idea), but as for whether it's an accurate portrayal or not I don't know. But regardless, dwindling in numbers and living in a world where big archosaurs were now controlling all the terrestrial ecosystems those poor last dicynodonts gradually faded into oblivion, until they were but dust in the wind dude. That was a Bill and Ted reference.

* There are several different views on how the term Mammalia should be defined (Rowe & Gauthier 1992). Here I've decided to follow what seems to be the most widely favoured view, and include all the Triassic mammal-like cynodonts (including *Adelobasileus*, morganucodontids and so on) within Mammalia.

In June 1915 several fragmentary fossil bones were discovered near Hughenden in Queensland (Australia). Heber A. Longman (best known for his 1924 description of the giant pliosaur *Kronosaurus*) exhibited them at a meeting in 1915, and noted that they resembled dicynodont elements. Well, it turns out that he was right, as a 2003 reappraisal of the specimens by Tony Thulborn and Susan Turner showed that the bones could not belong to anything *other than* a dicynodont. One of the most telling of the specimens is a partial maxilla that still houses its slightly recurved canine tusk (Fig. 95). In every detail – the distribution of concavities and foramina, the articulatory surfaces for other bones, the tooth shape, wear pattern and surface microstructure, the internal tooth structure (determined by CT scanning) – the specimen is indisputably dicynodont, and not matched by anything else.

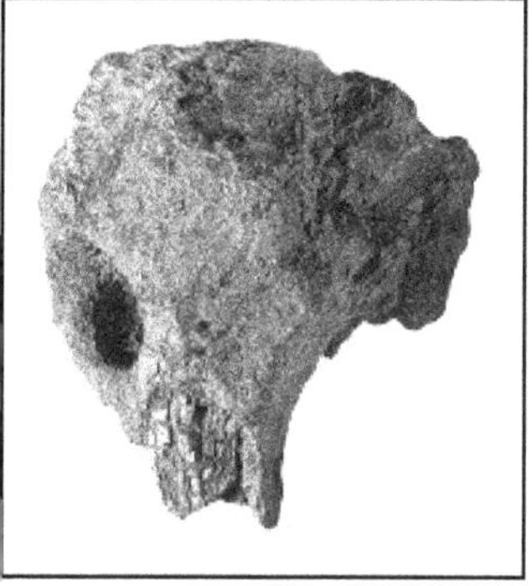

Fig. 95. Not the prettiest fossil in the world: the chunk of dicynodont maxilla, from Thulborn & Turner (2003).

But here's the big deal: the fossils are from the late Early Cretaceous, and thus something like 100 million years younger than the previously known youngest members of the group. Thulborn & Turner (2003) noted that this "is so extraordinary than it demands exceptionally rigorous investigation" (p. 987), and they carefully showed why and how other groups could be excluded from consideration. Furthermore, the Cretaceous age of the specimen is well established and there is no reason to doubt it. So dicynodonts didn't disappear in the Late Triassic as we'd always thought. They had in fact been sneakily surviving somewhere, and as Thulborn & Turner (2003) wrote, their persistence in Australia and absence from everywhere else suggests that "Australia's tetrapod fauna may have been as distinctive and anachronistic in the Mesozoic as it is at the present day" (p. 991). That's pretty incredible.

Refs - -

- King, G. 1990. *The Dicynodonts: A Study in Palaeobiology*. Chapman & Hall (London, New York).
- Maisch, M. W. 2001. Observations on Karoo and Gondwana vertebrates. Part 2: A new skull-reconstruction of *Stahleckeria potens* von Huene, 1935 (Dicynodontia, Middle Triassic) and reconsideration of kannemeyeriiform phylogeny. *Neues Jahrbuch fur Geologie und Palaontologie, Abhandlungen* 220, 127-152.
- Martill, D. M. & Naish, D. 2000. *Walking With Dinosaurs: The Evidence*. BBC Worldwide (London).
- Rowe, T. & Gauthier, J. 1992. Ancestry, paleontology, and definition of the name Mammalia. *Systematic Biology* 41, 372-378.
- Thulborn, T. & Turner, S. 2003. The last dicynodont: an Australian Cretaceous relict. *Proceedings of the Royal Society of London B* 270, 985-993.

Chapter 33:

'Time wandering' cynodonts and docodonts that (allegedly) didn't die: late-surviving synapsids II

We here continue with the subject initiated in the previous chapter: the amazing late survival of non-mammalian synapsids. By the way, the title I previously used ('late-surviving non-mammalian synapsids') no longer applies as some of the taxa discussed here are not non-mammalian synapsids, but are in fact basal mammals. We do begin with alleged non-mammalian taxa, however.

The last non-mammalian synapsid?

The most controversial of these 'Mesozoic survivors' is a diminutive and enigmatic animal from Upper Palaeocene North America: *Chronoperates paradoxus*, described by Fox *et al.* (1992) for a partial mandible and some isolated teeth from the Paskapoo Formation of Alberta. With a lower jaw that (when complete) would have been less than 30 mm long, and with tooth crowns about 2 mm tall,

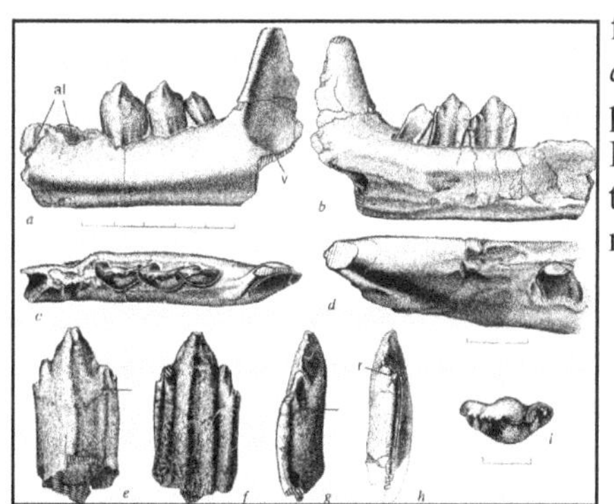

Fig. 96. The type specimen of *Chronoperates paradoxus*, as figured by Fox *et al.* (1992). Diagrams a-d show the dentary fragment and e-i show an isolated tooth. The scale bar divisions correspond to 1 mm.

Chronoperates would have been shrew-sized. While clearly synapsid in identity, Fox *et al.* (1992) noted that several features made this taxon amazingly archaic.

Firstly, the medial surface of the dentary possesses trough-like concavities that Fox *et al.* interpreted as areas for the articulation of post-dentary bones. Crown-group mammals lack these, but they were of course widespread in other synapsids. Secondly, the teeth are unusually tall and laterally compressed, and with an archaic multi-cusped crown morphology. Superficially, the teeth resemble those of Triassic non-mammalian cynodonts like *Therioherpeton*. Thirdly, the tooth enamel of *Chronoperates* is reportedly of pseudoprismatic type, a morphology not present in placental mammals and thus by inference expected in non-mammalian synapsids. Among synapsids, this combination of features is reportedly only seen in non-mammalian cynodonts, and Fox *et al.* concluded that *Chronoperates* must have been one of these: the first of them (at the time) to come from post-Jurassic strata. If this is correct, then the geological range of non-mammalian cynodonts had just been extended by about 100 million years, and it's for this reason that the name *Chronoperates* means 'wanderer through time' (since 1992, Upper Jurassic and Cretaceous non-mammalian cynodonts have been described, thereby shortening this gap).

All in all, the case doesn't sound too bad and initial reports were quite positive: wow, mammals weren't the only synapsids that survived into the Cenozoic after all, and non-mammalian synapsids had been there all along, hiding in the proverbial shadows (Novacek 1992). Fox and his colleagues have stuck to their guns, and in recent publications have continued to regard *Chronoperates* as a non-mammalian cynodont (Scott *et al.* 2002). How was it that post-Jurassic basal cynodonts had remained undetected for so long? Fox *et al.* proposed that the small size of these specimens, their rarity, their possible restriction to northerly regions, and a lack of the right kind of sampling, might explain their previous absence from the fossil record.

Their proposal was soon criticized. In a rebuttal, Hans Sues (1992) argued that all of the features employed by Fox *et al.* (1992) to demonstrate non-mammalian status for *Chronoperates* were problematical. Sues argued that the teeth of non-mammalian cynodonts were actually pretty different from those of *Chronoperates*, that Mesozoic mammals didn't share pseudoprismatic enamel with *Chronoperates*, and that the interpretation of the troughs on the dentary as probable attachment areas for post-dentary bones was unconvincing. He concluded that "the fossils currently available do not justify classification of *Chronoperates* as a non-mammalian cynodont" (p. 278). Furthermore, while Fox *et al.* (1992) were careful to exclude other possible contenders (including crocodyliforms, pterosaurs, lizards, xenarthrans and mesonychians) there were a number of Palaeogene mammal groups that they didn't exclude, and it would have been more convincing had it been shown that these taxa could also be eliminated from comparison.

But if *Chronoperates* isn't what Fox *et al.* said it was, what is it? A few possibilities have been suggested, or hinted at, in the literature. In their *Classification of Mammal: Above the Species Level*, McKenna & Bell (1997) classified *Chronoperates* as a basal holotherian (Holotheria = kuehneotheriids, spalacotherioids, dryolestoids, therians etc.). They noted that its position is "dubious, even at this level, but not a nonmammalian cynodont" (p. 43). So, while they didn't agree with Fox *et al.* that *Chronoperates* was non-mammalian, they still thought that it was closest to taxa that were otherwise Triassic and Jurassic. Well, hey, that's about as exciting as the possibility that *Chronoperates* might be a non-mammalian synapsid. I mean: a relative of kuehneotheriids and gobiconodontids that survived into the Cenozoic? It would be a big deal.This possibility (that *Chronoperates* is a late-surviving basal holotherian mammal) has been hinted at by some other workers. In a study on the distribution of an ossified Meckel's cartilage in basal mammals, Meng *et al.* (2003) noted that the medial dentary scar seen in *Chronoperates* might not house post-dentary bones, as Fox *et al.* proposed, but instead a persisting Meckel's cartilage. Now, if *Chronoperates* did possess a Meckel's cartilage, this would be a first for a post-Mesozoic synapsid, and would further support ideas that *Chronoperates* is actually a late-surviving basal mammal. And as I noted above, this would be very exciting. For now, this is all we know of *Chronoperates*, and it remains a controversial enigma, and possibly an exciting anachronism.

Last of the docodonts?

So whatever it is, *Chronoperates* remains enigmatic and poorly known. Quite the opposite is true of the docodonts, a relatively successful and long-lived group of basal mammals that were long known only from the Late Jurassic North American form *Docodon* Marsh, 1881. Docodonts had characteristically broad molars and a complex dentition that appears to have been rather like that of tribosphenic mammals, and both the homology of their tooth cusps, and their affinities to other mammals, have proved controversial. In their recent phylogenetic review of Mesozoic mammals, Luo *et al.* (2002) found docodonts to be way down near the base of Mammalia "despite their precociously specialized dentition" (p. 16). Most other workers seem to agree with this.

Fig. 97. Life restorations of docodonts are, understandably, somewhat rare. This life restoration of the amphibious docodont *Castorocauda* (named in 2006 from the Daohugou Beds of China) was produced by Arthur Weasley.

Early Cretaceous docodonts weren't described until 1928 when G. G. Simpson reported *Peraiocynodon inexpectatus* from the Purbeck Limestone of England. *Peraiocynodon* is quite similar to *Docodon* and a number of workers have regarded the two as congeneric. Since the 1970s, several additional docodont taxa have been described from the British Middle Jurassic (see Sigogneau-Russell 2003 for review) and Middle Jurassic forms are also known from Kyrgyzstan, China and Siberia. An Upper Jurassic form, *Haldanodon*, was described from Portugal in 1972, and in 1994 an Upper Jurassic Mongolian form, *Tegotherium*, was described. A second Early Cretaceous docodont, *Sibirotherium*, was named by Maschenko *et al.* (2002) for material from Siberia. *Sibirotherium* shares some characters with *Tegotherium*, and both taxa have been united in the clade Tegotheriidae. In its 2002 description, *Sibirotherium* was said to be the youngest reported member of the group.

But here we come to the surprise. Were docodonts all but a distant memory by the Late Cretaceous? Well, in 2000 Rosendo Pascual *et al.* claimed that, no, they weren't, but that they had actually managed to hang on until the very end of the Late Cretaceous. Pascual *et al.* (2000) described how the poorly known Cretaceous mammal *Reigitherium bunodontum*, originally described by Jose Bonaparte in 1990 as a dryolestoid, appeared in fact to be a bona fide docodont, and a very late surviving one.

Discovered in the Campanian-Maastrichtian La Colonia Formation of Patagonia, if *Reigitherium* were a docodont, it would show that they had survived almost to the end of the Cretaceous, and – if they'd been lucky – they might even have scraped through into the Cenozoic. Pascual *et al.*'s proposal would also extend the docodont record by a minimum of 30 million years given that *Sibirotherium* might be as young as Aptian-Albian. Alternatively, *Sibirotherium* might be as old as Berriasian-Valanginian, in which case *Reigitherium* would extend the docodont record by about 55 million years. Either way, it would be significant.

But is *Reigitherium* really a docodont? According to an abstract produced by Rougier *et al.* (2003), no it isn't, as new material from the La Colonia Formation later showed that *Reigitherium* really was a dryolestoid, as originally identified by Bonaparte. Technical details of the molar cusps and roots, and the tooth count, better match those of dryolestoids more than docodonts, and in fact *Reigitherium* appears particularly closely related to the Palaeocene dryolestoid *Peligrotherium*, with both being united by Rougier *et al.* (2003) in the dryolestoid clade Reigitheriidae. Oh well. I confess that I had totally missed this abstract and didn't know of it until David Marjanović pointed it out to me. The scepticism that some Mesozoic mammal workers had about the alleged docodont status of *Reigitherium* might explain why Maschenko *et al.* (2002) appeared to 'overlook' the 2000 reidentification of *Reigitherium* as a late-surviving docodont, hence their claim that *Sibirotherium* was the youngest known docodont.

Refs - -

- Fox, R. C., Youzwyshyn, G. P. & Krause, D. W. 1992. Post-Jurassic mammal-like reptile from the Palaeocene. *Nature* 358, 233-235.
- Luo, Z., Kielan-Jaworowska, Z. & Cifelli, R. L. 2002. In quest for a phylogeny of Mesozoic mammals. *Acta Palaeontologica Polonica* 47, 1-78.

- Maschenko, E. N., Lopatin, A. V. & Voronkevich, A. V. 2002. A new genus of tegotheriid docodonts (Docodonta, Tegotheriidae) from the Early Cretaceous of West Siberia. *Russian Journal of Theriology* 1, 75-81.
- McKenna, M. C. & Bell, S. K. 1997. *Classification of Mammals: Above the Species Level*. Columbia University Press (New York).
- Meng, J., Hu, Y., Wang, Y. & Li, C. 2003. The ossified Meckel's cartilage and internal groove in Mesozoic mammaliaforms: implications to origin of the definitive mammalian middle ear. *Zoological Journal of the Linnean Society* 138, 431-448.
- Novacek, M. J. 1992. Wandering across time. *Nature* 358, 192.
- Pascual, R., Goin, F. J., González, P., Ardolino, A. & Puerta, P. F. 2000. A highly derived docodont from the Patagonian Late Cretaceous: evolutionary implications for Gondwanan mammals. *Geodiversitas* 22, 395-414.
- Rougier, G., Novacek, M. J., Ortiz-Jaureguizar, E., Pol, D. & Puerta, P. 2003. Reinterpretation of *Reigitherium bunodontum* as a Reigitheriidae dryolestoid and the interrelationships of the South American dryolestoids. *Journal of Vertebrate Paleontology* 23 (Supp. 3), 90.
- Scott, C. S., Fox, R. C. & Youzwyshyn, G. P. 2002. New earliest Tiffanian (late Paleocene) mammals from Cochrane 2, southwestern Alberta, Canada. *Acta Palaeontologica Polonica* 47, 691-704.
- Sigogneau-Russell, D. 2003. Docodonts from the British Mesozoic. *Acta Palaeontologica Polonica* 48, 357-374.
- Sues, H.-D. 1992. No Palaeocene 'mammal-like reptile'. *Nature* 359, 278.

Chapter 34:
Make that ten most 'beautifully interesting' birds

A while back I was tagged to perpetuate something known as the '10 bird meme'. The idea here is that you list your ten 'most beautiful birds'. As you'll know if you've heard of this project, it was started by John at A DC Birding Blog and has so far produced over 50 responses. To be honest I'm not that interested in or excited by the human concept of what is considered 'beautiful', so I've added my own slant to this and have decided to cover instead a selection of birds that can be considered 'beautiful' in terms of what they tell us about evolution. We might call them 'beautifully interesting' birds. Given that there are about 10,000 extant bird species to chose from, my selection is pretty much random and hardly representative. I simply sat down and wrote about the first ten 'beautifully interesting' species that popped into my head. Then I started writing about them... Here we go, I hope you enjoy.

1. Flying steamer-duck *Tachyeres patachonicus*

Admittedly, all ten of my most 'beautifully interesting' birds could be anseriforms, as I have a special affinity for waterfowl. But I'll limit myself to one of my favourites, the Flying steamer-duck.

The most widely distributed of the four *Tachyeres* species (one of which, *T. leucocephalus*, was only described in 1981), *T. patachonicus* inhabits both the fresh and marine waters of the Falklands and southern Patagonia and Tierra del Fuego. While all other steamer-ducks are flightless, *T. patachonicus* is (obviously) not, and in contrast to its flightless relatives it has proportionally bigger pectoral muscles and lower wing loadings. But what makes the species especially interesting is that some males within the species actually have wing loadings that are too high to permit flight, and are thus flightless (Humphrey & Livezey 1982, Livezey & Humphrey 1986). So, within a single species, there are both flighted and flightless individuals. It is almost as if the species is poised in the transition to full flightlessness, and indeed both morphological and genetic studies (Corbin *et al.* 1988) agree that *T. patachonicus* is the most basal member of its otherwise flightless genus. Flighted and flightless individuals are known to have also occurred in some recently extinct anseriform species, incidentally.

Fig. 98. Female Falkland steamer-duck *Tachyeres brachypterus*.

But there's more. Steamer-ducks are notoriously pugnacious. Heavy-bodied and robust compared to other ducks, they have tough skin, a massive head and neck, and are equipped with keratanised orange knobs on the proximal parts of their carpometacarpi. Both sexes use these wing knobs in territorial fights and displays. Fighting males grab each other by the head or neck and then whack each other vigorously with the wing knobs, and fights can last for up to 20 minutes. Both birds sometimes submerge during the fight, and come up still fighting. This reminds me of scenes in films where super-heroes and villains (e.g., Spider-Man vs Doc Ock) fall off buildings together and continue to battle even while plummeting toward the ground, but that's just me. An aggressive steamer-duck approaches an 'enemy' by either adopting the so-called submerged sneak posture (only the top of the head and back and tail tip are visible), or by 'steaming' noisily across the surface (the ducks charge at speed, throwing their wings like the paddles of a paddle-steamer, hence the vernacular name).

Here's where things get especially cool. Other waterbirds are shit-scared of steamer-ducks, and 'mass spooks' of other duck species, grebes and coots have been recorded when these birds saw or heard the local *T. patachonicus*. You see, they attack and kill other waterfowl. A particularly detailed steamer-duck attack on a Shoveler *Anas platalea* was recorded by Nuechterlein & Storer (1985a), and I here summarise the account they describe on p. 89.

A male steamer-duck caught a male shoveler by the neck and began pounding it with its wing knobs. The female steamer-duck displayed excitedly nearby. The shoveler was held beneath the water, then yanked up and beat some more. The male steamer-duck took a break and displayed with his female, then he went back to the shoveler, grabbed it again by the neck and proceeded to beat it another 15-20 times. By now the shoveler was looking pretty limp (though still alive). It was pecked at and released and both steamer-ducks displayed together again, and the male steamer-duck now began to move away from the shoveler. The shoveler now began to move (slowly) toward the shore and eventually got there. Then it died. "Examination of the specimen disclosed several broken bones, hemorrhages in the lower neck region and massive internal bleeding at the base of the right leg" (p. 89). During the course of their study at Laguna de la Nevada, Santa Cruz Province, Argentina, Nuechterlein & Storer (1985a) picked up the carcasses of six ducks that had definitely been killed by steamer-ducks within a single week. Why steamer-ducks are so aggressive remains the source of debate (Murray 1985, Livezey & Humphrey 1985a, b, Nuechterlein & Storer 1985b). But don't mess with them.

2. Shovel-billed kingfisher *Clytoceyx rex*

A New Guinea endemic, *C. rex* was named in 1880 and it's been studied by such ornithological luminaries as Forshaw, Beehler and Diamond. A halcyonid kingfisher and close relative (or even member) of the kookaburra group, it's strikingly big (c. 30 cm long) and with a fantastic broad, flattened bill. This is driven hard into the ground with a vigorous action and it's apparently used quite literally as a shovel. As is the case in some other birds that sometimes handle hard-shelled prey, the edges of its tomia are scopate: that is, they possess tiny brush-like structures (Gosner 1993). Poorly known and rarely seen, it apparently does most of its foraging at dusk. Mostly preying on worms, the species also eats lizards, snails and insects. It's sometimes called the Earthworm-eating kingfisher. It's cool.

Fig. 99. There are hardly any good photos of the Shovel-billed kingfisher, so I had to knock this drawing up.

3. Wrybill *Anarhynchus frontalis*

Tetrapods generally have symmetrical bodies, but there are a few awesome exceptions. Among birds, the best is the Wrybill, a mid-sized charadriid wader endemic to New Zealand. Its short bill, about 25 mm long, is remarkable and unique in being curved at its tip toward the right, and always to the right. To forage, the bird tilts its head to the left and explores beneath stones and rocks, scraping off

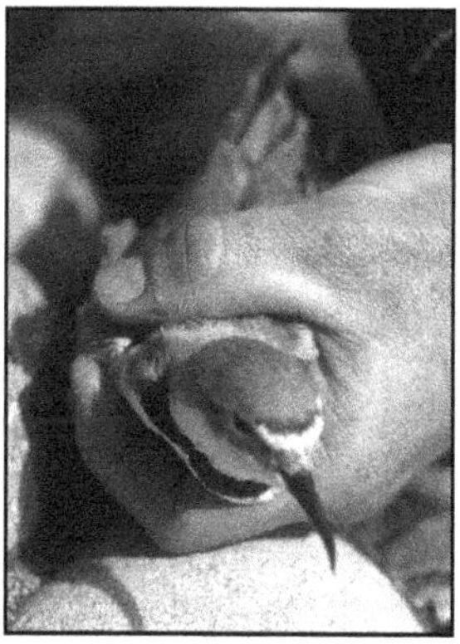

Fig. 100. A female Wrybill, photo by John Hill.

arthropods, fish eggs and other objects. It always moves clockwise around a stone. In *The Life of Birds*, David Attenborough suggested that this unique behaviour might only have been able to evolve in a place where terrestrial predators were absent: the Wrybill keeps watch for aerial predators while foraging, but it doesn't need to watch for terrestrial ones. Apart from its bill, the Wrybill isn't much to look at. It's grey and white with a black breast band and white forehead. Phylogenetic studies find it to be close to *Charadrius* (Chu 1995), and thus probably a core charadriid (traditional Charadriidae is not monophyletic (Ericson *et al.* 2003)). Wrybills migrate from North Island to South Island to breed. The total world population is only 4000-5000.

4. Blue-capped ifrita *Ifrita kowaldi*

Originally described by DeVis in 1890 as *Todopsis kowaldi*, *Ifrita* was independently 'discovered' by Walter Rothschild in 1898 and named by him *Ifrita coronata*. A passerine endemic to moist montane forests on New Guinea, *Ifrita* is remarkable for two reasons. Firstly, nobody really knows what it is and over the years it's been classified in several different, disparate passerine families. It's been allied with warblers, log-runners, and corvids. Secondly, it's poisonous. I'll repeat that for those people who hadn't heard it before. It's poisonous. While it's nowadays reasonably well known that pitohuis (a group of six species of pachycephalid passerines, also endemic to New Guinea) produce batrachotoxin in their skin and feathers, it was shown in 2000 that *Ifrita* does too (Dumbacher *et al.* 2000). It's thought that the poisons present in these birds are sequested from poisonous insect prey, but last I heard this was still under debate.

As for why the birds are poisonous, it's been widely suggested that the poisons they harbour function as a chemical defence against snakes, raptors and predatory mammals. However, they may also protect the birds against parasites (Mouritsen & Madsen 1994). Incidentally, (1) it seems that not all pitohui species are poisonous (although further study is required to be absolutely sure about this), (2) that another New Guinean passerine, the Rufous shrike-thrush *Colluricincla megarhyncha*, also produces batrachotoxin, and (3) that multiple other non-poisonous New Guinea passerines (including some other pitohuis) may mimic poisonous pitohuis and therefore gain protection from predators too (Diamond 1992, Dumbacher & Fleischer 2001).

5. Ground tit *Pseudopodoces humilis*

If convergence is one of the most interesting evolutionary phenomena, then the Ground tit should become a text-book example of it, on par with thylacines vs wolves and ichthyosaurs vs dolphins. Described in 1871 by A. Hume, the Ground tit is a weak-flying brown passerine of the Tibetan plateau, often superficially likened to a wheatear. But for most of the time that we've known of it, it has not gone by the name Ground tit at all: rather, it has been termed Hume's ground-jay (or Little ground-jay or Tibetan ground-jay or Hume's ground-pecker). This is because, you see, it was always regarded as a ground-jay, that is, as a terrestrial corvid. While superficially similar to true ground-jays (the four *Podoces* species), it was always regarded as a highly aberrant member of Corvidae, and as the smallest member of the group.

Fig. 101. Tibetan groundpecker *Pseudopodoces humilis*.

Hume in fact initially described *P. humilis* as a member of *Podoces*. Like *Podoces*, *P. humilis* possesses a slender, decurved bill, pale brown plumage and a dry, open-country habitat. However, they're also highly different. While ground-jays run, *P. humilis* hops, and while ground-jays use stick nests, *P. humilis* nests in tunnels or burrows. Ground-jays are also much larger than *P. humilis* and exhibit white wing patches and dark, iridescent plumage patches. In recognition of these differences, *P. humilis* was given its own subgenus within *Podoces* in 1902, and in 1928 this was elevated to generic status.

But the allocation of *P. humilis* to Corvidae wasn't really doubted until prominent osteological differences between *P. humilis* and ground-jays were noted by Borecky (1978). Borecky doubted the classification of *P. humilis* as a corvid and hinted at an affinity with starlings. In her 1989 PhD study on corvid phylogeny, Sylvia Hope agreed that *P. humilis* was utterly unlike corvids, and most like nuthatches and tits. Despite these objections, *P. humilis* has remained classified as a corvid in most standard works on Corvidae (Goodwin 1986, Madge & Burn 1999) and indeed in most general works on birds. To resolve the issue once and for all, Helen James and colleagues performed a detailed analysis of the morphology and genetics of *P. humilis*, comparing it widely with other passerines (James *et al.* 2003). All the data showed, pretty conclusively, that *P. humilis* is not a corvid, but in fact a parid. A tit. A unique, highly novel tit to be sure, but a tit nonetheless, hence the new vernacular name.

6. Blakiston's fish owl *Bubo blakistoni*

If you think evolutionary convergences are cool, then you'll love reversals. Morphological features or aspects of behaviour that have been modified during the evolution of a lineage don't have to 'stay' changed – as Louis Dollo thought they did (this is where so-called Dollo's Law come from) – they can change back to the ancestral state if this is what works. Of the four fish owl species, Blakiston's is the biggest and most formidable, with a wingspan approaching 2 m and a total length of c. 60 cm. Endemic to Siberia, eastern China, Japan, Sakhalin Island and other eastern Asian islands, it inhabits cool, remote forests and can cope with harsh winters. Some have suggested that during its history it may have suffered from competition with sea eagles (Hume 1991), which are similar in size and ecological requirements. Sadly, this remarkable bird is highly endangered.

As is well known, owls in general have soft plumage and unusual fringes of tiny barbs along the leading margins of their flight feathers. These features – both specializations that permit silent flight in a group of birds that rely on sensitive hearing – are derived relative to the condition that owls inherited from their ancestors. But fish owls don't need to be silent given that

Fig. 102. Blakiston's fish owl, photographed at Kushiro Zoological Park, Japan.

they've specialized to prey on animals that live under water, and they've consequently reversed back to the primitive condition. Fish owls also have longer legs than those of other owls, and their feet sport rough spiny scales that resemble those of fish-eating raptors like ospreys. *B. blakistoni* is unique among fish owls in having feathered legs. Interestingly, fish owls walk down to the water's edge and will even wade into the shallows. They then sit motionless, waiting for prey to come within range. They don't just eat aquatic prey, but also terrestrial birds and mammals. Fish owls are also remarkable among owls in reportedly feeding on carrion.

How does *B. blakistoni* fit into owl phylogeny? Until recently, the fish owls were considered to represent a distinct genus, *Ketupa*, and *Ketupa* was considered closely related to, but distinct from, the eagle owls *Bubo*. Recent genetic studies have found instead that the *Ketupa* species are nested within *Bubo* (as is *Nyctea*, the Snowy owl), and consequently both *Ketupa* and *Nyctea* have been sunk into synonomy with *Bubo* (Wink & Heidrich 1999). These results are supported by osteological characters, but unfortunately this data has yet to be published (it's included in Ford's 1967 PhD thesis, and I've heard that a version of this is due to be published soon). I don't have a copy of König *et al.* to hand, so I don't know exactly how the fish owl species fit into *Bubo*. The feathered legs and other characters of *B. blakistoni*, however, suggest that, among fish owls, this species is the most basal. Overall, it seems like the one fish owl that is most like 'normal' eagle owls.

7. Shoebill *Balaeniceps rex*

Also called the Shoe-billed stork, She-billed stork [not a typo], Whale-bill or Whale-headed stork, *B. rex* is a long-legged big-billed waterbird of central Africa, and a specialist denizen of papyrus swamps. Though known to the ancient Egyptians, it wasn't described by science until John Gould named it in 1851. Before that time it was a cryptid, as an 1840 sighting of this as-of-then-unidentified bird had been published by Ferdinand Werne in 1849 (Shuker 1991).

Standing 1.4 m tall, the Shoebill can exceed 2.6 m in wingspan and is best known for its remarkable wide bill. This can be up to 25 cm long, is larger in males than females and, like that of pelicans, cormorants and gannets, lacks external nostril openings. The birds use the bill to grab at large aquatic prey like lungfishes, catfish, tilapia, snakes, turtles and frogs. They're reputed to eat antelope calves, but this is highly unlikely to say the least (Renson 1998), and apparently carrion. Little known is that the Shoebill is one of a handful of birds that occasionally practices quadrupedality: when Shoebills lunge forward while grabbing prey, they sometimes use their wings to help push themselves upright.

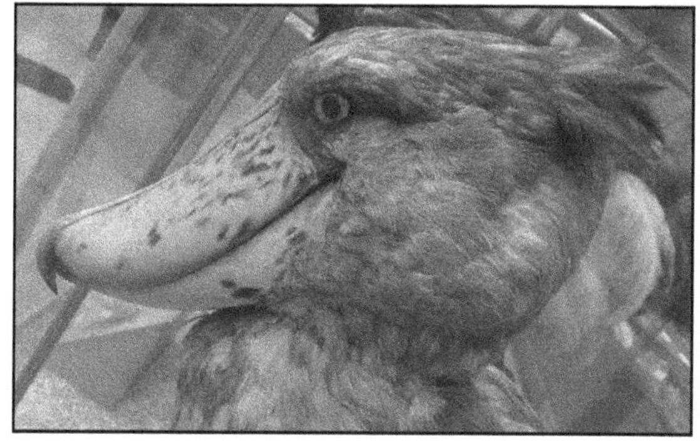

Fig. 103. Head of a stuffed Shoebill.

The affinities of the Shoebill have been controversial. Gould regarded it as a pelican and data from egg-shell microstructure and ear morphology was used by later authors to support this view. Unlike pelicans however, the long toes of the Shoebill are unwebbed and it is stork-like in some aspects of behaviour, practicing bill clattering and also dribbling water onto its eggs and young during the heat of the day. Based on stapedial morphology, Feduccia (1977) argued that the Shoebill really is a stork. It is also heron like in its possession of powder-down and some other features, and some workers have argued that it is really an aberrant heron. As recently shown by Gerald Mayr (2003) however, the morphological evidence best supports a position for the Shoebill close to Steganopodes, the clade that includes frigatebirds, pelicans, gannets, cormorants and anhingas (traditional Pelecaniformes is not monophyletic as tropicbirds are apparently closer to procellariiforms than they are to members of Steganopodes).

8. African harrier-hawk *Polyboroides typus**

Sometimes called the Gymnogene, *Polyboroides* is a gracile, naked-faced raptor with grey and black plumage, but best known for the so-called double-jointedness present in its intertarsal joints. The foot can actually bend both anteriorly and posteriorly (that is, the foot can be hyperextended as well as flexed), which needless to say is (almost) unique (read on). This is easy to see in dead or anaesthetized specimens, but it's reportedly hard to observe in action in living birds. Cooper (1980) wrote that, after five months of observing captive specimens, the notebook was still being filled with such comments as "I can't really fathom what is the 'double-jointedness' they refer to in the books".

However, the behaviour has now been well documented and filmed, and the birds use this incredible flexibility to extricate lizards, insects and other prey from fissures in rocks, and nestling birds from their nests. You might presume, as I did, that *Polyboroides* has a unique sort of intertarsal joint, perhaps with the trochlear surfaces of the distal tibiotarsus wrapping onto the posterior surface of the bone as well as the anterior surface. But according to Cooper (1980) "in anatomical structure there is no significant difference between [the intertarsal joints of *Polyboroides*] and the corresponding joints of Kestrel, Tawny eagle or Black kite" (p. 98). Err, gosh. I haven't yet checked to see if there are any more recent studies of harrier-hawk ankles (let me know if you know better), but I find that pretty amazing. As I like to say sometimes 'Anatomy is not destiny' (not my quote, I stole it from a paper on armadillos).

* I'm only discussing *P. typus* here, but there is another extant species in the genus (Madagascan *P. radiatus*).

Fig. 104. African harrier-hawk, as illustrated by the great natural history illustrator Louis Agassiz Fuertes in 1930.

Polyboroides is 'beautifully interesting' for two other reasons. Firstly, it's amazingly similar to a South American raptor, the Crane hawk *Geranospiza caerulescens*. *Geranospiza* looks pretty much the same as *Polyboroides*, occupies the same ecological niche, behaves in the same manner, and even has the same bizarrely mobile intertarsal joint. Understandably, the two genera have often been regarded as each other's closest relatives (e.g., Brown 1997). But.... they're also different in many subtle anatomical details (Burton 1978), and consequently there's been a long-standing debate as to whether they're related or not. While unfortunately few phylogenetic studies include both species together, a comprehensive recent DNA-based phylogeny found the two to be well apart, with *Polyboroides* down with Old World vultures and honey buzzards, while *Geranospiza* was in the buteonine clade that also includes *Buteo* (of course) and *Leucopternus* (Lerner *et al.* 2005). This study is pretty compelling and the phylogeny near-conclusive. *Polyboroides* and *Geranospiza* thus represent striking, amazing instances of convergent evolution, one of the best examples of this among birds.

Incidentally, there's a second raptor that also evolved striking convergence with *Polyboroides*, but it's a fossil form and I'll have to talk about it another time.

Secondly, *Polyboroides* is interesting for being one of the most basal members of Accipitridae (the hawk-eagle-Old World vulture family). As hinted at in the previous paragraph, morphological and DNA-based phylogenies of Accipitridae tend to agree that one of the most basal clades in the group is that which includes Old World vultures and honey buzzards. Recent studies agree that *Polyboroides* is in this clade, and within it one of the most basal members (Holdaway 1994, Lerner *et al.* 2005). That makes it one of the most basal members of the whole of the accipitrid radiation. So you have there a raptor that (1) does something anatomically freaky, (2) does that anatomically freaky thing employing a mechanism that no-one yet properly understands, (3) exhibits uncanny, striking convergence with an unrelated raptor on another continent, and (4) is one of the most archaic, phylogenetically basal members of its group.

9. Kakapo *Strigops habroptilis*

"... [it] clambers up and down trees because it cannot fly ... it purrs like a cat and smell like a posy of fragrant flowers … it allows itself to be picked up and handled without demur or apparent concern" (Vietmeyer 1992, p. 69).

Sometimes called the Owl parrot, the Kakapo was first described by John Gray in 1845. Everything about kakapos is extraordinary. Remember that it's a parrot while you read the following. A large nocturnal (!), cryptically-coloured (!!) terrestrial (!!!) bird, endemic to New Zealand, the kakapo is a specialized foliage-eater that seems to live on a metabolic knife-edge, rather like the Giant panda. Successful breeding is limited only to those years when mass fruitings of podocarps occur. This worked fine in a New Zealand where there were lots of podocarps and lots of kakapos (and indeed fossils show that kakapos were formerly abundant), but environmental destruction caused by humans meant that kakapos were forced into suboptimal areas where life was even harder than before. Combined with this was devastating predation from domestic cats, stoats and rats. Consequently, as is well known, kakapos have become extinct on the mainland and only survive on managed offshore islands where introduced rats and other predators have been eliminated (Vietmeyer 1992, Clout 2001). In 2001 there were 62 individuals. Worldwide.

Fig. 105. Kakapo, photographed on Codfish Island, New Zealand.

Apparently in order to bulk-process the poor-quality vegetation it eats, the Kakapo has evolved voluminous guts, and accordingly a larger overall body size. It has hence become a giant among parrots (with big males reaching 3.6 kg), and is the largest extant species. Conventionally stated to be flightless, it is in fact capable of gliding, so this is not strictly true. Certainly it mostly walks places however, and individuals create well-worn trails in the mountainous forest where they live. Distinctively 'chewed', compressed fragments of vegetation hang from the plants adjacent to these trails, and kakapos leave both compact cylindrical droppings and white traces of uric acid on the trails. The uric acid streaks apparently "have a herb-like smell when fresh" (Juniper & Parr 1998, p. 372).

Uniquely among parrots, kakapos are lek breeders, with males booming out loud calls from suitable topographical hollows, and these calls can be heard from about 1 km away. Cryptic plumage and nocturnal habits don't make much sense in an environment devoid of predators, so the fact that kakapos possess these traits suggests that there were once predators able to kill them. Now we know that New Zealand did possess such predators: endemic giant eagles and large harriers (see Chapter 1).

10. Eurasian oystercatcher *Haematopus ostralegus*

Finally, one of my most favourite birds is the extraordinary, charismatic, beautifully interesting oystercatcher. One of ten extant haematopodid species, it sports pied plumage, pinkish legs, and has the heaviest bill of any extant wader (Fig. 106). One of the most interesting things about oystercatchers is the fact that they exhibit resource polymorphism, with some populations exhibiting multiple different forms (Skúlason & Smith 1995). 'Stabbers' feed by jabbing their laterally compressed bill tips in between the valves of a mussel's shell, while 'hammerers' crack open mussel shells by pounding on them. Some hammerers only break in to the shell on its dorsal side, while others only break in to the ventral side. Others attack only the left side valve, and others only the right valve. Others are worm specialists with pointed tweezer-like bill tips.

First discovered by M. Norton-Griffiths during the 1960s (and extensively studied by a great many ornithologists since then), resource polymorphism among oystercatchers was initially thought to be learnt by the birds from their parents (and not genetically determined). It now seems that things are far more flexible, with individuals switching from one behaviour to the other over the years. It's been said that juveniles can't really learn how to handle prey from their parents given that many of them are reared inland and are abandoned by their parents before they ever get to the coast (Sutherland 1987). However, some oystercatcher adults spend up to a year teaching their young how to exploit prey: in fact a photo in Attenborough's *The Life of Birds* shows an adult opening a shell while a juvenile, at its side, watches with apparent interest.

It seems that it's the behavioural flexibility that controls bill shape, rather than the other way round, and another remarkable thing about oystercatchers is how specialized their bills are for coping with wear. Uniquely among waders, the bill grows at a jaw-dropping 0.4 mm per day (that's three times faster than the growth rate of human fingernails). This rapid growth means that the bill can change shape very rapidly if the feeding style is changed, and captive individuals that were forced to switch from bivalve-feeding to a diet of lugworms changed from having chisel-shaped bills to tweezer-like bills within 10 days. A-maz-ing.

Fig. 106. Eurasian oystercatcher, photo by Neil Phillips.

Given that oystercatchers are fairly large and powerful for waders, and able to smash open bivalve shells, it follows that they are formidable and potentially dangerous to other birds. Certainly males will chase off raptors when defending nesting females. I recall reading accounts of them caving in the heads of other waders during territorial disputes, but unfortunately I can't remember where (a common problem, despite my well organized library). Most aggressive interactions recorded between oystercatchers, and between oystercatchers and other waders, involve piracy, and in fact some birds obtain most of their food this way, "attacking other birds at an average o five minute intervals during low tide" (Hammond & Pearson 1994, p. 61). As much as 60% of the food of some individuals is obtained by piracy. Finally, oystercatchers are incredibly long-lived, with the record-holder dying at age 35! Now, come on, that is a truly extraordinary bird.

Refs - -

- Borecky, S. R. 1978. Evidence for the removal of *Pseupodoces humilis* from the Corvidae. *Bulletin of the British Ornithologists's Club* 98, 36-37.
- Brown, L. H. 1997. *Birds of Prey*. Chancellor Press, London.
- Burton, P. J. K. 1978. The intertarsal joint of the harrier-hawks *Polyboroides* spp. and the Crane hawk *Geranospiza caerulescens*. *Ibis* 120, 171-177.
- Chu, P. C. 1995. Phylogenetic reanalysis of Strauch's osteological data set for the Charadriiformes. *The Condor* 97, 174-196.
- Clout, M. 2001. Where protection is not enough: active conservation in New Zealand. *Trends in Ecology & Evolution* 16, 415-416.
- Cooper, J. E. 1980. Additional observations on the intertarsal joint of the African harrier-hawk *Polyboroides typus*. *Ibis* 122, 94-98.
- Corbin, K. W., Livezey, B. C. & Humphrey, P. S. 1988. Genetic differentiation among steamer-ducks (Anatidae: *Tachyeres*): an electrophoretic analysis. *The Condor* 90, 773-781.
- Diamond, J. M. 1992. Rubbish birds are poisonous. *Nature* 360, 19-20.
- Dumbacher, J. P. & Fleischer, R. C. 2001. Phylogenetic evidence for colour pattern convergence in toxic pitohuis: Müllerian mimicry in birds? *Proceedings of the Royal Society of London B* 268, 1971-1976.
- Dumbacher, J. P., Spande, T. F. & Daly, J. W. 2000. Batrachotoxin alkaloids from passerine birds: a second toxic bird genus (*Ifrita kowaldi*) from New Guinea. *Proceedings of the National Academy of Sciences, USA* 97, 12970-12975.
- Ericson, P. G. P., Envall, I., Irestadt, M. & Norman, J. A. 2003. Inter-familial relationships of the shorebirds (Aves: Charadriiformes) based on nuclear DNA sequence data. *BMC Evolutionary Biology* 3: 16.
- Feduccia, A. 1977. The whalebill is a stork. *Nature* 266, 719-720.
- Goodwin, D. 1986. *Crows of the World*. Trustees of the British Museum (Natural History) (London).
- Gosner, K. L. 1993. Scopate tomia: an adaptation for handling hard-shelled prey? *Wilson Bulletin* 105, 316-324.
- Hammond, N. & Pearson, B. 1994. *Waders*. Hamlyn, London.
- Holdaway, R. N. 1994. An exploratory phylogenetic analysis of the genera of the Accipitridae, with notes on the biogeography of the family. In Meyburg, B.-U. & Chancellor, R. D. (eds) *Raptor Conservation Today*. WWGBP/The Pica Press, pp. 601-649.
- Hume, R. 1991. *Owls of the World*. Parkgate Books (London).
- Humphrey, P. S. & Livezey, B. C. 1982. Flightlessness in flying steamer-ducks. *The Auk* 99, 368-372.
- James, H. F., Ericson, P. G. P., Slikas, B., Lei, F.-M., Gill, F. B. & Olson, S. L. 2003.

Pseudopodoces humilis, a misclassified terrestrial tit (Paridae) of the Tibetan Plateau: evolutionary consequences of shifting adaptive zones. *Ibis* 145, 185-202.

- Juniper, T. & Parr, M. 1998. *Parrots*. Pica Press, Mountfield.
- Lerner, H. R. L. & Mindell, D. P. 2005. Phylogeny of eagles, Old World vultures, and other Accipitridae based on nuclear and mitochondrial DNA. *Molecular Phylogenetics and Evolution* 37, 327-346.
- Livezey, B. C. & Humphrey, P. S. 1985a. Territoriality and interspecific aggression in steamer-ducks. *The Condor* 87, 154-157.
- Livezey, B. C. & Humphrey, P. S. 1985b. Interspecific aggression in steamer-ducks. *The Condor* 87, 567-568.
- Livezey, B. C. & Humphrey, P. S. 1986. Flightlessness in steamer-ducks (Anatidae: *Tachyeres*): its morphological bases and probable evolution. *Evolution* 40, 540-558.
- Madge, S. & Burn, H. 1999. *Crows & Jays*. Christopher Helm (London).
- Mayr, G. 2003. The phylogenetic affinities of the Shoebill (*Balaeniceps rex*). *Journal of Ornithology* 144, 157-175.
- Mouritsen, K. N. & Madsen, J. 1994. Toxic birds: defence against parasites? *Oikos* 69, 357-358.
- Murray, B. G. 1985. Interspecific aggression in steamer-ducks. *The Condor* 87, 567.
- Nuechterlein, G. L. & Storer, R. W. 1985a. Aggressive behavior and interspecific killing by Flying steamer-ducks in Argentina. *The Condor* 87, 87-91.
- Nuechterlein, G. L. & Storer, R. W. 1985b. Interspecific aggression in steamer-ducks. *The Condor* 87, 568.
- Renson, G. 1998. The bill. *BBC Wildlife* 16 (10), 10-18.
- Shuker, K. P. N. 1991. *Extraordinary Animals Worldwide*. Robert Hale (London).
- Skúlason, S. & Smith, T. B. 1995. Resource polymorphisms in vertebrates. *Trends in Ecology and Evolution* 10, 366-370.
- Sutherland, W. J. 1987. Why do animals specialize? *Nature* 325, 483-484.
- Sutherland, W. J. 2002. Science, sex and the kakapo. *Nature* 419, 265-266.
- Vietmeyer, N. D. 1992. The salvation islands. In Calhoun, D. (ed) *1993 Yearbook of Science and the Future*. Encyclopaedia Brittanica Inc (Chicago), pp. 60-75.
- Wink, M. & Heidrich, P. 1999. Molecular evolution and systematics of the owls (Strigiformes). In König, C., Weick, F. & Becking, J.-H. *Owls: a Guide to the Owls of the World*. Pica Press (London), pp. 39-57.

Chapter 35:
Introducing the plethodontids

What are plethodontids? Often called lungless salamanders (for the obvious reason), they're a predominantly American clade, consisting of over 280 species of terrestrial, aquatic, cave-dwelling and arboreal salamanders. They range in length from 40 to 325 mm: that upper limit is a pretty respectable size for a salamander, but they presumably can't get much bigger than this for reasons of surface area : volume ratio constraints. Conventionally allied with salamandrids, plethodontids may in fact be closest to amphiumas and rhyacotritonids. Uniquely among salamanders, many of them have evolved direct development: that is, the eggs hatch into miniature adults, and the larval phase has been skipped (read on for more on this).

They are also odd among salamanders in that some species can drop the tail as a predator-defence mechanism (properly called caudal autotomy), and in that some species have only four toes on the hindfeet. Some plethodontids curl up into a ball and roll downhill to escape predators (García-París & Deban 1995), a behaviour reported elsewhere in tetrapods among pangolins by the way (Tenaza 1975). While many species capture insects with a projectile tongue, *Speleomantes supramontis* (one of the European plethodontids) actually fires its tongue skeleton out of its mouth, with the tongue protruding for up to 80% of its body length during this act. It's "the only vertebrate known to shoot part of the visceral skeleton completely out of its body as a projectile" (Deban *et al.* 1997, p. 28).

North America has a high diversity of plethodontids, with 14 genera and c. 120 species. South America is home to about eight genera (some of which also occur in North America) and c. 150 species. Note that, for reasons that I will discuss elsewhere, these figures are vague and hard to verify. One genus and one genus alone, *Speleomantes*, the cave salamanders, is present in Europe. *Speleomantes* Dubois, 1984 was previously included within *Hydromantes* Gistel, 1848, and whether the two should be considered distinct remains one of the most contested areas in plethodontid systematics (and a new name, *Hydromantoides* Lanza & Vanni, 1981, has been used by some authors for the American members of *Hydromantes*).

Fig. 107. Sardinian cave salamander *Speleomantes genei*, one of seven species conventionally included in *Speleomantes*. Photo by Franco Andreone.

If *Hydromantes* and *Speleomantes* are the same genus, the distribution of this taxon is pretty odd: it occurs as relict populations in Italy, France and Sardinia, and then also in California. If these populations do belong to the same genus, then this is a pretty impressive example of so-called discontinuous distribution. The North American *Hydromantes* species are sometimes called web-toed salamanders.

When I was learning the herpetofauna of Europe as a teenager, *Hydromantes* (= *Speleomantes*) was always said to include just two species (*H. italicus* Dunn, 1923 and *H. genei* (Temminck & Schegel, 1838)). Recent studies have led to the resurrection or naming of a further five species (*H. ambrosii* Lanza, 1955, *H. strinatii* Aellen, 1958, *H. imperialis* Stefani, 1969, *H. supramontis* Lanza *et al.*, 1986 and *H. flavus* Stefani, 1969), all of which can be distinguished morphologically (Griffiths 1995). Inhabitants of caves and rocky outcrops where there is running water, they're pretty bizarre, with small sensory tentacles on the upper lip, an almost immobile lower jaw, an oval glandular swelling on the chin, partially webbed toes and a semi-prehensile tail. They're good climbers and able to ascend even vertical surfaces, and *H. imperialis* "emits a strong smell" apparently. We've already seen above the amazing projectile tongue some species possess.

Note that all of the European cave salamander species listed above, bar one, were named in the 20th century. New plethodontid species are routinely described from tropical America as well as from North America, including from well-studied 'non-remote' areas. *Batrachoseps campi*, described for the first time by Wake (1996), is from the San Gabriel Mountains: a location less than 50 km from downtown Los Angeles. A quick look at author dates for plethodontids reveals numerous recently named genera, among them *Bradytriton* Wake & Elias, 1983, *Dendrotriton* Wake & Elias, 1983, *Nototriton* Wake & Elias, 1983, *Nyctanolis* Wake & Elias, 1983, *Ixalotriton* Wake & Johnson, 1989, and *Cryptotriton* García-París & Wake, 2000. Some of these are long-known but have only recently been split from other genera, while others really are new discoveries. Incidentally, the 'Wake' listed there is David B. Wake, a leading expert on plethodontids at the University of California.

Fig. 108. Cerro de Enmedio moss salamander *Nototriton lignicola*, a plethodontid named in 1997. Photo by Josiah H. Townsend.

The presence in Europe of plethodontids is unusual enough (albeit well known), but the biggest surprise in plethodontid research has been the recent discovery of an Asian member of the group, the Korean crevice salamander *Karsenia koreana* Min *et al.*, 2005 (Fig. 109). First collected in 2003, *Karsenia* is a small, somewhat robust plethodontid that recalls *Plethodon* in overall appearance (Min *et al.* 2005). Its discovery indicates a long, and hitherto totally unknown, history of plethodontids in continental Asia, and the fact that only one Asian species is known suggests that rates of speciation were far lower among Asian plethodontids than they were among American ones. What next among the Asian herpetofauna – extant albanerpetontids?*

All of this data on new species matches the species discovery rates recorded for other lissamphibian groups. A 13% increase in valid recognised lissamphibians occurred between 1985 and 1992, with a further 6% increase from 1992 to 1995. Glaw & Köhler (1998) correctly predicted that the extant number of valid lissamphibian species would exceed 5000 by the year 2000, and Hanken (1999) wondered "Why are there so many new amphibian species when amphibians are declining?".

The phylogenetic arrangement of plethodontid clades is interesting. In the classic morphology-based phylogeny produced by Wake (1966), a desmognathine clade is sister to a far larger

* For those who don't know, that was a reference to the recent discovery of albanerpetontids in Pliocene strata (Venczel & Gardner 2005).

plethodontine clade that includes Hemidactyliini* (c. 25 species with aquatic larvae, including the genera *Eurycea*, *Hemidactylium* and so on), Bolitoglossini (c. 220 species lacking aquatic larvae, including the genera *Bolitoglossa*, *Dendrotriton*, *Oedipina*, *Pseudoeurycea* and so on) and Plethodontini (c. 55 species lacking aquatic larvae: the genera *Aneides*, *Ensatina* and *Plethodon*).

However, a recent molecular analysis (Mueller *et al.* 2004) rejected the monophyly of ALL of these groups with the exception of Desmognathinae (but rather than being the sister-taxon to Plethodontinae it was nested within this group, and within it was found to be the sister-taxon to a '*Hydromantes*' + *Aneides* clade). If this new phylogeny is valid (I should note by the way that Wake was on the authorship of Mueller *et al.*, so it's not as if this study disagrees with his own research), it means that we have to re-assess the evolution of the different life history strategies seen in plethodontids. Wake argued that direct development had evolved twice in plethodontids (once within desmognathines and once again at the base of the bolitoglossin + plethodontin clade), and in both cases it had evolved from ancestors that possessed a larval stage.

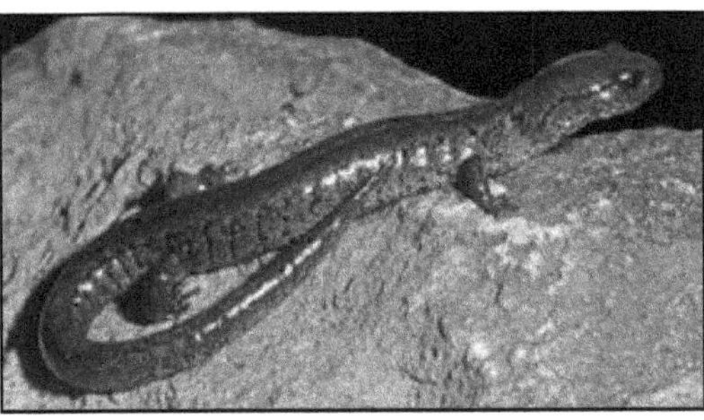

Fig. 109. Korean crevice salamander *Karsenia koreana*. Photo by David Vieites of the University of California, Berkeley.

But Mueller *et al.*'s (2004) cladogram favours more complex scenarios. They argued that at least four transitions from the possession of a biphasic life history to the presence of direct development had occurred. Furthermore, in at least one instance, the presence of a larval phase must have re-evolved from an ancestor with direct development: "a morphological transition rarely reported and previously considered unlikely" (Mueller *et al.* 2004, p. 13823). That's pretty incredible, but it's perhaps not without precedent among lissamphibians as there are some indications that one group of marsupial tree-frogs may also have switched from direct development back to the possession of a tadpole stage.

This area was also considered by Chippindale *et al.* (2004): their phylogeny wasn't as heterodox as that of Mueller *et al.* (2004), but they did agree on the recovery of desmognathines within Plethodontinae. Because desmognathines possess a larval stage, while the taxa around them in the new phylogeny are all direct-developing species, desmognathines simply must have re-evolved biphasic life history from direct development. So, as Chippindale *et al.* (2004) discussed, the reversal of direct development in plethodontine plethodontids shows yet again that complex features may be regained (they drew analogy with the re-evolution of wings in stick insects and of hindlimbs in snakes, though the evidence for the latter is still controver-

* This name was coined by Wake in 1966, and some have argued that it is a synonym of Mycetoglossini Bonaparte, 1850.

sial).

Why re-evolve the larval stage however? Surely it's good to stick with direct development once it's evolved? Well, is it a coincidence that desmognathines occur in an area where plethodontid diversity is extremely high (with up to 11 direct-developing species occurring sympatrically), where non-desmognathine plethodontids dominate all ecosystems available to salamanders except for stream habitats, and in an area where direct-developing plethodontids are (get this) THE most important vertebrates in terms of biomass (Burton & Likens 1975)? Chippindale *et al.* (2004) therefore made a good case that desmognathines have re-evolved biphasic life history in order to exploit the 'only' available habitats, those represented by streams and streamsides, in regions densely packed with other plethodontid species. Desmognathines have clearly done well at this, sometimes occurring at densities of 6.9 individuals per square metre, and even evolving species that prey on direct-developing plethodontines.

Refs - -

- Burton, T. M. & Likens, G. E. 1975. Salamander populations and biomass in the Hubbard Brook Experimental Forest, New Hampshire. *Copeia* 1975, 541-546.
- Chippindale, P. T., Bonett, R. M., Baldwin, A. S. & Wiens, J. J. 2004. Phylogenetic evidence for a major reversal of life-history evolution in plethodontid salamanders. *Evolution* 58, 2809-2822.
- Deban, S. M., Wake, D. B. & Roth, G. 1997. Salamander with a balistic tongue. *Nature* 389, 27-28.
- García-París, M. & Deban, S. M. 1995. A novel antipredator mechanism in salamanders: rolling escape in *Hydromantes platycephalus*. *Journal of Herpetology* 29, 149-151.
- Glaw, F. & Köhler, J. 1998. Amphibian species diversity exceeds that of mammals. *Herpetology Review* 29, 11-12.
- Griffiths, R. A. 1996. *Newts and Salamanders of Europe*. T & A D Poyser, London.
- Hanken, J. 1999. Why are there so many new amphibian species when amphibians are declining? *Trends in Ecology & Evolution* 14, 7-8.
- Min, M. S., Yang, S. Y., Bonett, R. M., Vieites, D. R., Brandon, R. A. & Wake, D. B. 2005. Discovery of the first Asian plethodontid salamander. *Nature* 435, 87-90.
- Tenaza, R. R. 1975. Pangolins rolling away from predation risks. *Journal of Mammalogy* 56, 257.
- Venczel, M. & Gardner, J. D. 2005. The geologically youngest albanerpetontid amphibian, from the Lower Pliocene of Hungary. *Palaeontology* 48, 1273-1300.
- Wake, D. B. 1966. Comparative osteology and evolution of the lungless salamanders, family Plethodontidae. *Memoirs of the South California Academy of Sciences* 4, 1-111.
- Wake, D. B. 1996. A new species of *Batrachoseps* (Amphibia: Plethodontidae) from the San Gabriel Mountains, southern California. *Contributions in Sciences of the Natural History Museum of Los Angeles County* 463, 1-12.

Chapter 36:

The interesting and contentious discovery of the kipunji, and new light on the evolution of drills, mandrills and baboons

Monkeys are one of the most misunderstood and mischaracterized of all mammals, at least among those who aren't primate specialists. 'Monkey' is itself a loose term used for those members of Anthropoidea that aren't hominoids, and here we're only interested in the Old World monkeys (cercopithecoids). Pet peeve no. 375 is the fact that people think that apes are somehow superior to monkeys in morphology and adaptability. If anything it's the other way round, with a phenomenal post-Miocene radiation of monkeys pushing the previously hyper-diverse apes into the shadows. That's a story for another time, however. One monkey in particular has been the focus of a lot of attention lately, the Highland mangabey *Lophocebus kipunji* (although this is not its current name, as we'll see shortly).

Fig. 110. The monkey originally known as the Highland mangabey *Lophocebus kipunji*.

The story of the Highland mangabey's discovery is an interesting one. In January 2003 Tim Davenport of the Tanzanian Wildlife Conservation Society "heard rumours [from the local Wanyakyusa people of the Mount Rungwe region] about a shy and atypical monkey known as Kipunji" (Beckman 2005, Jones *et al.* 2005, p. 1161), and became interesting in tracking down the species that might lay at the bottom of these reports. Meanwhile, another primatologist – Trevor Jones – had been amazed to observe an unusual, unidentifiable monkey in the Tanzanian Ndundulu Forest Reserve, a location about 350 km away from the source of the Kipunji reports.

So, at this stage, we have an ethnoknown primate known only to scientists by way of fleeting observations. This makes the Kipunji a bona fide cryptid, and, to repeat a point I've made before (in connection with the Odedi: see Chapter 26), one would be justified in arguing that Davenport, Wood and their colleagues were now engaging in cryptozoological research. By definition these primatologists are therefore part-time cryptozoologists, whether they like it or not.

Good observations were finally made of the Kipunji in December 2003, and it now turned out that a monkey population reported in 2001 from the Ndundulu Forest Reserve, misidentified as Sanje mangabeys *Cercocebus sanjei* (itself only discovered in 1979), were actually reports of the Kipunji. Davenport's team and Wood's team both learnt that they'd been tracking down two different populations of the same monkey in October 2004 – one population in the Mount Rungwe-Livingstone Forest area in the Southern Highlands of Tanzania, and the other in the Ndundulu Forest Reserve in the Tanzanian Udzungwa Mountains.

In May 2005 they published a jointly authored description of this new species in *Science* (Jones *et al.* 2005), naming it the Highland mangabey *Lophocebus kipunji* Ehardt, Butynski, Jones & Davenport, 2005. Note that the authorship of the species doesn't match the authorship of the descriptive paper. This does happen occasionally in the literature, and in this case it means that three authors of the original description were not involved in the naming of the species. Something else makes the Jones *et al.* paper odd, however, and this is the apparent lack of a type specimen for *Lophocebus kipunji*. Nowhere in the paper do they list or cite an accessioned specimen, nor mention the procurement of a specimen for a museum or university collection. Under 'holotype' they wrote "Adult male in photograph (Fig. 2)", with a 'paratype' being described as "Adult in photograph (Fig. 3)" (p. 1162). Because the new species seems to be critically endangered, Jones *et al.* (2005, p. 1162) stated "no live individual should be collected at this time to serve as the holotype".

While it might seem ethically 'nice' and – from the conservationist perspective – sensible to not collect specimens from endangered or declining populations, it's problematic to not do so, as the ICZN mandates that actual type specimens are needed for the establishment of a species. Furthermore, biological entities can only be regarded as unquestionably valid when there are physical specimens accessioned in proper collections.

Consequently there were several follow-up comments in *Science* on the lack of a Highland mangabey holotype (Landry 2005, Moser 2005, Polaszek *et al.* 2005, Timm *et al.* 2005). Timm *et al.* (2005) argued that the lack of a type specimen means that *Lophocebus kipunji* "is not an available name and has no formal standing in zoology" (p. 2163) and Landry (2005) argued that the authors should have published "all of the excellent descriptive material and their quite convincing case for calling it new, without, however, naming it" (p. 2164). Landry also noted that only four of the seven authors of the paper are listed as namers of the new species, but that "the purpose of this citation is to identify the paper, not to assign credit, and all of the authors should be cited". Interesting.

Polaszek *et al.* (2005) responded to these criticisms by arguing that the ICZN does actually allow the recognition of taxa without holotypes (Article 73.1.4), and they argued that conservation concerns should encourage zoologists to catalogue species on the basis of vocalisations, molecular information and so on, and that "dead animal specimens should not be understood to be essential to the process of establishing new taxa" (p. 2165). Hell, if Bernard Heuvelmans were still alive he would dance a little jig (to paraphrase Greg Paul). Moser (2005) pointed out that Polaszek *et al.* (2005) seem to have completely misunderstood the point of Article 73.1.4: it refers to cases where type specimens have become lost, and does NOT say that lack of an original type specimen is acceptable.

Anyway, it isn't the first time that this has happened, and it won't be the last. Among tetrapods the most famous case is that of the Bulo Burti boubou *Laniarius liberatus*, a Somalian shrike captured, observed in captivity, and then released (Smith *et al.* 1991). Exactly the same thing has just happened with another new monkey: a Brazilian platyrrhine named the Blond capuchin *Cebus quierozi* Mendes Pontes & Malta, 2006 (again, the species' authorship doesn't match the authorship of the paper. Again, weird). Apparently critically endangered and restricted to a tiny area already renowned as a centre of endemism, the authors elected not to sacrifice the type specimen (ironically confiscated from a local hunter), but to release it back into the wild (Mendes Pontes *et al.* 2006). I could talk about this subject a lot more, but won't do so here. Suffice it to say, I still think that the collection of specimens is an integral part of zoological science (see Patterson 2002). Naming species on the basis of photos, vocalizations and molecular data alone opens the door for the official recognition of taxa that *most* zoologists are not ready to accept: on these grounds, Sasquatch is clearly a valid, nameable taxon for example.

To get back to the Highland mangabey, the problematic lack of a type specimen is now an academic argument, as a museum-accessioned specimen has been procured: it is a sub-adult male found dead in a trap in August 2005, currently accessioned at the Field Museum of Natural History, Chicago. It possesses the diagnostic features of the specimens previously described as representing this species, so seems to be securely and correctly identified (note here one of the practical problems of lacking a holotype: we don't have a physical specimen that we can compare latterly obtained specimens to). The Chicago specimen is significant, as study of its DNA sequence data has provided new information on the affinities and evolution of the Highland mangabey, and on mangabeys and their relatives as a whole.

Kipunjis, magabeys, drills and mandrills

Mangabeys are an entirely African assemblage of cercopithecid monkeys, traditionally grouped together in the genus *Cercocebus* Geoffroy Saint-Hilaire, 1812. They're all superficially alike, being long-tailed and long-limbed, and with moderately long muzzles and large incisors (Fig. 111). However, molecular studies have consistently found mangabeys to be diphyletic, with the six terrestrial mangabey species forming a clade with drills and mandrills (and with macaques too in some studies), and the two arboreal mangabeys forming a clade with baboons and geladas.

Fig. 111. One of the mangabeys of the 'drill-magabey' group: Collared mangabey *Cercocebus torquatus*. Photo by B. S. Thurner Hof.

The two kinds of mangabeys also differ from one another in many morphological details. Consequently, it has been widely agreed that the two arboreal species should be split from *Cercocebus* and given their own genus, *Lophocebus* (originally coined by Palmer in 1903 to replace *Semnocebus* Gray, 1870. The latter needed replacing as it was preoccupied by *Semnocebus* Lesson, 1840, a name now regarded as a junior synonym of *Avahi* Jourdan, 1834). I like Jonathan Kingdon's use of the term 'drill-mangabey' for the terrestrial *Cercocebus* species, and of 'baboon-mangabey' for the arboreal *Lophocebus* species, and I'll use them from hereon. Incidentally, drill-mangabeys are sometimes called eyelid monkeys because of their white upper eyelids, and baboon-mangabeys are sometimes called black mangabeys, for the obvious reason. This division of the mangabeys has been mostly accepted by mammalogists, but not universally so (McKenna & Bell 1997 still treat all mangabeys as the single genus *Cercocebus*, for example).

If correct, this diphyletic take on mangabey affinities would mean that geladas, baboons, drills and mandrills do not form a clade of 'dog-faced cercopithecids' as conventionally thought.

Apparently good morphological support for the non-monophyly of mangabeys came from Fleagle & McGraw's (1999) study. They found that drill-mangabeys and drills and mandrills shared numerous features that aren't present in baboon-mangabeys and baboons. In the humerus, drill-mangabeys, drills and mandrills share a notably broad deltoid plane, a proximally extended supinator crest, a broad flange for the brachialis, and a narrow olecranon process with a deep lateral ridge, and there are also characters in the radius and ulna that unite these monkeys to the exclusion of their close relatives. Drill-mangabeys, drills and mandrills are also united by particularly large, rounded posterior premolars, a robust ilium, a reduced gluteal tuberosity on the femur, sharp borders to the margins of the patellar groove, and other characters.

So that's a pretty impressive list of characters, but note that they're all associated with the terrestrial foraging style that these monkeys employ. Drill-mangabeys, drills and mandrills search manually through rotten wood and leaf litter, consuming hard nuts and seeds, and audibly cracking them with their large teeth. Furthermore, outgroup comparison (with macaques for example) indicates that some of these characters are primitive for the cercopithecid clade that includes these species (Papionina). If we only had this morphological data, the possible monophyly of a drill-mangabey + drill/mandrill clade would be suspicious (Fleagle & McGraw 1999). But of course we don't just have this morphological data, we also have the molecular data discussed above. So things are looking pretty robust.

And this is where the new data from that Chicago specimen of the Highland mangabey comes in. Its genetic sequences independently confirm the relationships indicated by previous molecular studies, and by Fleagle & McGraw's morphological characters. Though initially described as a third species of baboon-mangabey (Jones *et al.* 2005), the Highland mangabey's DNA show instead that it is closer to baboons that it is to baboon-mangabeys, yet it lacks the characters that unite all baboons proper. It therefore needs to be recognized as a new genus, and accordingly it's now known as *Rungwecebus kipunji* Davenport *et al.*, 2006. These authors proposed that members of *Rungwecebus* should now be referred to as kipunjis, and not as mangabeys anymore. Within Papionina, *Rungwecebus* and *Papio* form a clade, and the *Rungwecebus* + *Papio* clade forms a trichotomy with baboon-mangabeys and geladas. The sister-group to this kipunji/baboon/gelada/baboon-mangabey clade is the drill-mangabey + drill/mandrill clade (Davenport *et al.* 2006).

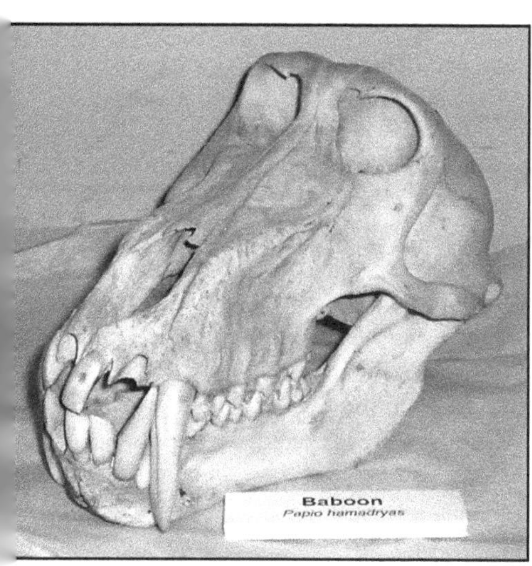

Fig. 112. A baboon skull. This skull was labelled as that of a Hamadryas baboon *Papio hamadryas* but is more likely to be that of a Chacma baboon *P. ursinus*.

Given that we now have dog-faced monkeys variously scattered about a phylogenetic tree that also includes shorter-faced baboon-mangabeys, drill-mangabeys and the kipunji, you should now be wondering about the polarity and evolution of the long muzzle and large body size seen in mandrills, baboons and geladas. Is the long muzzle (Fig. 112) and large size primitive for this clade, or have these features evolved convergently two or even three times? At the moment, we can't be sure: if the long muzzle evolved at the base of Papionina, three reversals to the short-snouted condition must have occurred, but if basal members of Papionina were short-snouted, you need either two or three independent acquisitions of the long-snouted condition. Neither scenario is clearly more parsimonious than the other.

Kingdon (1997) argued that the skull morphology of the drill-mangabeys indicates that they are dwarfed, short-faced descendants of large drill-like ancestors. Do fossils help? They might, as there are assorted fossil papionins, some of which (like *Pliopapio*) are long-snouted, others of which (like *Parapapio*) are relatively short-snouted (Frost 2001, Leakey *et al.* 2003). Unfortunately the phylogenetic affinities of these fossil taxa remain contentious.

What's nice about this whole mangabey-kipunji-baboon subject is that it illustrates a point I've been planning to make for a while: namely, that fossils are sometimes all but useless in determining evolutionary relationships, and we most certainly don't *require* them in order to uncover phylogenetic patterns. Contrary to what lay-people (and creationists) seem to think, you do not *need* fossils in order to uncover and reveal the reality of evolution, and even if fossils didn't exist, scientific logic would lead us to conclude that organisms changed over time.

Don't get me wrong, fossils are great (err, I am a palaeontologist if I remember correctly), and they can certainly elucidate things and inform us about evolution and past diversity, but it's living organisms, not dead ones, that provide the best evidence for evolutionary change.

Refs - -

- Beckman, M. 2005. Biologists find new species of African monkey (*Lophocebus kipunji*). *Science* 308, 1103.
- Davenport, T. R. B., Stanley, W. T., Sargis, E. J., De Luca, D. W., Mpunga, N. E., Machaga, S. J. & Olson, L. E. 2006. A new genus of African monkey, *Rungwecebus*: morphology, ecology, and molecular phylogenetics. *Sciencexpress* 10.1126/science.1125631
- Fleagle, J. G. & McGraw, W. S. 1999. Skeletal and dental morphology supports diphyletic origin of baboons and mandrills. *Proceedings of the National Academy of Science* 96, 1157-1161.
- Frost, S. R. 2001. New early Pliocene Cercopithecidae (Mammalia: Primates) from Aramis, Middle Awash Valley, Ethiopia. *American Museum Novitates* 3350, 1-36.
- Jones, T., Ehardt, C. L., Butynski, T. M., Davenport, T. R. B., Mpunga, N. E., Machaga, S. J. & De Luca, D. W. 2005. The Highland mangabey *Lophocebus kipunji*: a new species of African monkey. *Science* 308, 1161-1164.
- Kingdon, J. 1997. *The Kingdon Field Guide to African Mammals*. Academic Press (San Diego).
- Landry, S. O. 2005. What constitutes a proper description? *Science* 309, 2164.
- Leakey, M. G., Teaford, M. F. & Ward, C. V. 2003. Cercopithecidae from Lothagam. In Leakey, M. G. & Harris, J. M. (eds) *Lothagam: the Dawn of Humanity in Eastern Africa*. Columbia University Press (New York), pp. 201-248.
- Mendes Pontes, A. R., Malta, A. & Henrique Asfora, P. 2006. A new species of capuchin monkey, genus *Cebus* Erxleben (Cebidae, Primates): found at the very brink of extinction in the Pernambuco Endemism Centre. *Zootaxa* 1200, 1-12.
- McKenna, M. C. & Bell, S. K. 1997. *Classification of Mammals: Above the Species Level*. Columbia University Press, New York.
- Moser, M. 2005. Holotypic ink. Science e-letters http://www.sciencemag.org/cgi/eletters/309/5744/2163c
- Patterson, B. D. 2002. On the continuing need for scientific collecting of mammals. *Journal of Neotropical Mammalogy* 9, 253-262.
- Polaszek, A., Grubb, P., Groves, C., Ehardt, C. L. & Butynski, T. M. 2005. What constitutes a proper description? Response. *Science* 309, 2164-2166.

- Smith, E. F. G., Arctander, P., Fjeldså, J. & Amir, O. G. 1991. A new species of shrike (Laniidae: *Laniarius*) from Somalia, verified by DNA sequence data from the only known individual. *Ibis* 133, 227-235.
- Timm, R. M., Ramey, R. R. & The Nomenclature Committee of the American Society of Mammalogists. 2005. What constitutes a proper description? *Science* 309, 2163-2166.

Chapter 37:

Chewed bones and bird-eating microbats

Megadermatids, the false vampires or yellow-winged bats, are a group of tetrapod-eating bats of the Old World and Australasian tropics. The best known members are the Australian false vampire or Ghost bat *Macroderma gigas* and the two Asian false vampires of the genus *Megaderma*. Long thought to be blood-feeders, megadermatids are odd in lacking upper incisors, and in having strongly reduced, thread-like premaxillae. They're big, with *Macroderma* weighing up to 150 g and having a wingspan of 60 cm. Nycterids, the mostly African slit-faced bats, will also prey on tetrapods (frogs, birds and smaller bats). Phyllostomids, the New World leaf-nosed bats, include the tetrapod-eating genera *Vampyrum*, *Trachops* (famous for being a specialist predator of singing frogs) and *Chrotopterus*. They're superficially like megadermatids, and even bigger, with Linnaeus' false vampire *Vampyrum spectrum* having a wingspan of about 1 m (making it the largest microbat).

A few other tetrapod-eating bats can be found in other microbat clades, like Vespertilionidae. A few pteropodids (fruit bats), notably the Hammer-headed fruit bat *Hypsignathus montrosus*, have been reported to feed from carrion and even attack tethered chickens (Hill & Smith 1984), and the New Zealand mystacinids are also known to sometimes scavenge from carcasses.

Fig. 113. Ghost bat *Macroderma gigas,* photographed at Perth Zoo with an infrared camera.

The foraging and predation techniques used by megadermatids have been reasonably well studied. They do eat invertebrates, but a proportion of their diet is made up of mammals (including rodents and other bats) and birds, and they also eat frogs and fish. Of all bats, megadermatids perhaps have the most flexible foraging style. They use low intensity, broad-band FM echolocation calls, but also listen acutely with their massive ears for prey-generated noises, and they may also hunt prey by sight. Altringham (1999) drew attention to this possibility, noting that the eyes of some species "are almost owl-like" (p. 219).

Using steath and darkness while hunting, they fly slowly among trees and shrubs and rocky areas, often close to the ground. Their attack on a tetrapod is culminated as the bat drops onto the prey, encases it in its wings and bites it hard on the head or neck (Kulzer *et al.* 1984). The prey is then carried back to the roost where it is eaten, and they can carry prey weighing about 60% of their own body weight. Mammal prey are usually eaten entirely, with just bits of the skull and parts of the gut, legs and tail being dropped to the floor. The situation with birds is somewhat different however, and went mostly unstudied until Walter Boles (an Australian palaeornithologist) produced a detailed paper on bird predation as practiced by the Ghost bat *Macroderma gigas* (Boles 1999).

The documented bird prey of *Macroderma* includes some 50+ species, most of which are ground-foraging passerines. Owlet-nightjars seem to be important prey items. In contrast to what happens with mammal prey, birds aren't eaten whole. Damage to the posterior part of the sternum shows that the bats eat tissue from the abdominal region, and leave the area around the anterior part of the sternum and the coracoids alone. The distal parts of the wings and legs get dropped intact and undamaged, as do remiges and rectrices. Humeri seem to get chewed up and eaten (Boles 1999).

Given that Boles is a palaeornithologist, you should now be wondering why a specialist on fossil birds was so interested in the behaviour of an extant bat species. One of the most famous Cenozoic fossil sites in the world is Riversleigh in north-west Queensland (see Archer *et al.* 1996 for outstanding coverage of the whole fauna). Numerous bird fossils are known from Riversleigh, including ratites, dromornithids, storks, rails, raptors*, parrots, kingfishers, swifts and passerines (including logrunners and lyrebirds).

The bat fauna discovered at Riversleigh is also diverse: literally millions (according to Archer *et al.* 1996) of bat bones are known from there, representing (as of 1996) "more than 35 different kinds of bats" (p. 135). Interestingly, at least five of them are megadermatids. And what's *really* interesting is that many of the Riversleigh bird bones represent those same skeletal elements that megadermatids drop from the bird carcasses they eat, and possess damage matching that caused by megadermatids. It therefore seems that megadermatids were important accumulators of the avian remains discovered at Riversleigh (Boles 1999, p. 88).

It's well known that owls are important accumulators of vertebrate remains, particularly in

* *Pengana robertbolesi*, a Riversleigh raptor named by Boles (after his father), is yet another raptor that convergently evolved a hyper-mobile tarsal joint like that of *Polyboroides* (Boles 1993).

caves, but this is the first study showing that bats play this role as well. Given that, as discussed above, there are other tetrapod-eating bats elsewhere in the world, it would be interesting to know if these species also produce piles of bones beneath their roosts. I wonder if anyone has looked.

Megadermatids are well known as predators of smaller tetrapods, so no one would be really surprised to hear about any of the stuff I've just discussed above. But you might be surprised to learn, as I was, that an inoffensive little vespertilionid is also an awesome tetrapod predator. That's what we'll be looking at in the next chapter.

Refs - -

- Altringham, J. D. 1999. *Bats: Biology and Behaviour*. Oxford University Press, Oxford.
- Archer, M., Hand, S. J. & Godthelp, H. 1996. *Riversleigh: The Story of Animals in Ancient Rainforests of Inland Australia*. Reed Books, Kew, Victoria.
- Boles, W. E. 1993. *Pengana robertbolesi*, a peculiar bird of prey from the Tertiary of Riversleigh, northwestern Queensland, Australia. *Alcheringa* 17, 19-25.
- Boles, W. E. 1999. Avian prey of the Australian ghost bat *Macroderma gigas* (Microchiroptera: Megadermatidae): prey characteristics and damage from predation. *Australian Zoologist* 31, 82-91.
- Hill, J. E. & Smith, J. D. 1984. *Bats: a Natural History*. British Museum (Natural History), London.
- Kulzer, E., Nelson, J. E., McKean, J. L. & Moehres, F. P. 1984. Prey-catching behaviour and echolocation in the Australian ghost bat, *Macroderma gigas* (Microchiroptera, Megadermatidae). *Australian Mammalogy* 7, 37-50.

Chapter 38:
Greater noctules: specialist predators of migrating passerines

In the previous chapter we looked at bird predation in the megadermatids (the false vampires or yellow-winged bats). Here, we're going to look at the bird predation recently documented in another bat group, the vespertilionids (vesper bats). To those of us living in the Northern Hemisphere, vesper bats are usually the most common and familiar of bats, and because they're generally small, inoffensive, insectivorous bats, the idea that some of them might be bird predators may seem pretty radical.

Noctules (*Nyctalus*) are a group of Eurasian vesper bats closely related to little yellow bats (*Rhogeesa*), big-eared brown bats (*Histiotus*) and little brown bats (*Myotis*) (Jones *et al.* 2002). Six species are currently recognized, although a few additional species are included by some experts in this genus as well (Nowak 1999). Mostly dwelling in forests, they are large (body lengths 50-100 mm, forearm lengths 40-70 mm) compared to pipistrelles and other typical vesper bats, and with sleek fur that ranges in colour from golden to dark brown. Some species are migratory and make journeys of over 2300 km. They eat relatively large prey, being particularly fond of beetles, and Nowak (1999) mentions a remarkable case where a Eurasian noctule (*N. noctula*) was observed to catch and eat mice. The species we're interested in here, the Greater noctule *N. lasiopterus*, is mostly western European but also occurs as far east as the Urals, and as far south as Libya and Morocco. It isn't well known, and is also rare.

To better understand the ecology and behaviour of this species, Ibáñez *et al.* (2001) netted individuals and recorded their wing morphology, and also recorded echolocation calls from the field. Greater noctule wing morphology indicates fast flight in open areas, as they have high wing loading and high aspect ratios. This is quite different from what's seen in the megadermatids and nycterids (see Chapter 37) where ground- and foliage-gleaning are employed to find and catch prey, and where prey are often hunted from perches. These bats use low wing loadings and low aspect ratios to practice slow, manoeuvrable flight as they glean for mostly terrestrial prey in cluttered habitats.

Fig. 114. A Greater noctule bares its teeth (this photo comes from Popa-Lisseanu *et al.* (2007): a key paper that appeared after my publication of this chapter's text).

The echolocation calls of the Greater noctule, being long in terms of pulse duration with a low frequency and narrow bandwidth, are suited for long-range target detection in open air. This technique is quite different from the prey detection style usually employed by vesper bats: a technique suited for short-range prey that are detected in cluttered habitats.

So, both wing morphology and echolocation style indicate that Greater noctules chase and catch flying prey in the open. But what are they chasing and catching? Ibáñez *et al.* (2001) found that feathers made up a significant component of the bat's droppings during March-May and again during August-November: those parts of the year when migrating birds pass through the Spanish study area. Both the proportion of feathers in the droppings, and the proportion of captured Greater noctules that produced feather-filled droppings, showed that Greater noctules must capture and eat large numbers of migrating passerines. Bird predation was first documented in this species by Dondini & Vergari (2000), and judging from later comments in the literature it doesn't seem that they got the credit they deserve for this discovery.

Unlike those of megadermatids, Greater noctule roosts are never littered with bird remains (which partly explains why this behaviour was overlooked for so long). Ibáñez *et al.* (2001) did find fresh, recently cut passerine wings (of a Robin *Erithacus rubecula* and Wood warbler *Phylloscopus sibilatrix*) underneath the bats they were netting, and they also found Robin feathers adhering to the claws of one of the captured bats. It seems that Greater noctules catch, overpower and eat the passerines during flight, just as other vesper bats do with flying insects. Greater noctules are clearly large enough and powerful enough to do this: they weigh in at about 50 g and have a wingspan of 45 cm (making them the largest of Europe's bats). For comparison, a European robin weighs c. 20 g (though this goes down to c. 15 g after migration) and has a wingspan of 20-22 cm.

Unsurprisingly, this idea has been regarded as controversial and doubtful by some. Bontadina & Arlettaz (2003) argued that the passerine-catching idea was so unlikely that it couldn't be regarded as correct (excellent use of logic there). They also noted that other noctule species prey on insects and not birds, and that, suspiciously, Greater noctule droppings lack bird bone fragments. None of these arguments count for much. Bird bone fragments in fact *are* present in Greater noctule droppings, as was reported by Dondini & Vergari (2000), who even subjected the bone fragments to SEM observation to determine conclusively their avian origin (strangely, Ibáñez *et al.* (2001) did not report the discovery of bone fragments, and Bontadina & Arlettaz (2003) didn't cite Dondini & Vergari's (2000) discovery of them). The fact that other noctules don't hunt passerines means nothing.

Anyway, how did Bontadina & Arlettaz (2003) account for the presence of all those feathers in the Greater noctule droppings? They proposed that the bats regularly eat falling feathers, mistaking them for flying insects! That's pretty incredible, and arguably more amazing than the idea of predation on passerines. Two responses to Bontadina & Arlettaz's article were published (Ibáñez *et al.* 2003, Dondini & Vergari 2004), and both showed that the scepticism was unfounded. The case for passerine predation in Greater noctules is pretty compelling. Note to wildlife camera-people reading this: someone should try and film this behaviour, though for obvious reasons no-one's even observed it yet (to my knowledge).

So Greater noctules are specialist predators that exploit nocturnally migrating passerines, and to date they are the only animals known to do so. There are diurnal raptors (notably Eleonora's falcon *Falco eleonorae* and Sooty falcons *F. concolor*) that specialize on migrating passerines, but nothing else that makes a point of catching those passerines that migrate at night. Dondini & Vergari's paper 'Carnivory in the greater noctule bat (*Nyctalus lasiopterus*) in Italy' and Ibáñez *et al.*'s paper 'Bat predation on nocturnally migrating birds' therefore have to be two of the most remarkable recent publications in the annals of bat science, and this is a field where lactation in males (Francis *et al.* 1994), ultraviolet vision (Winter *et al.* 2003) and the re-evolution of running (Riskin & Hermanson 2005) have recently been reported.

Of course bats don't have it all their own way against birds. Raptors and owls are important predators of bats and some studies suggest that owls may account for as much as 10% of annual bat mortality (Altringham 2003). Some raptors are bat-killing specialists, like the Bat kite (or Bat hawk) *Machaerhamphus alcinus* of tropical Africa, SE Asia and New Guinea, and the South American Bat falcon *Falco rufigularis*. In fact raptor predation seems to be so important to bats that it appears to explain why bats don't fly more during the daytime (Speakman 1991) and it may also explain why bats became nocturnal in the first place (Rydell & Speakman 1995). On the subject of raptors vs bats I was amused to see, in John Altringham's *Bats: Biology and Behaviour*, a drawing of a Harrier hawk *Polyboroides typus* using its flexible ankle joint to probe into a cavity where some molossid bats are roosting (p. 221) (for more on *Polyboroides* see Chapter 34).

Refs - -

Altringham, J. D. 2003. *British Bats*. HarperCollins, London.

Bontadina F. and Arlettaz R. 2003. A heap of feathers does not make a bat's diet. *Functional Ecology* 17, 141-142.

Dondini, G. & Vergari, S. 2000. Carnivory in the greater noctule bat *(Nyctalus lasiop-*

terus) in Italy. *Journal of Zoology* 251, 233-236.
- Dondini, G. & Vergari, S. 2004. Bats: bird-eaters or feather-eaters? A contribution to debate on Great noctule carnivory. *Hystrix* 15, 86-88.
- Francis, C. M., Anthony, E. L. P., Brunton, J. A. & Kunz, T. H. 1994. Lactation in male fruit bats. *Nature* 367, 691-692.
- Ibáñez, C., Juste, J., García-Mudarra, J. L. & Agirre-Mendi, P. T. 2001. Bat predation on nocturnally migrating birds. *Proceedings of the National Academy of Sciences* 98, 9700-9702.
- Ibáñez, C., Juste J., García-Mudarra J.L. & Agirre-Mendi P.T. 2003. Feathers as indicator of a bat's diet: a reply to Bontadina & Arlettaz. *Functional Ecology* 17, 143-145.
- Jones, K. E., Purvis, A., MacLarnon, A., Bininda-Emonds, O. R. P. & Simmons, N. B. 2002. A phylogenetic supertree of the bats (Mammalia: Chiroptera). *Biological Reviews* 77, 223-259.
- Nowak, R. M. 1999. *Walker's Mammals of the World, Sixth Edition.* The Johns Hopkins University Press, Baltimore and London.
- Popa-Lisseanu, A. G., Delgado-Huertas, A., Forero, M.G., Rodríguez, A., Arlettaz, R. & Ibáñez, C. 2007. Bats' conquest of a formidable foraging niche: the myriads of nocturnally migrating songbirds. *PLoS ONE* 2(2): e205. doi:10.1371/journal.pone.0000205
- Riskin, D. K. & Hermanson, J. W. 2005. Independent evolution of running in vampire bats. *Nature* 434, 292.
- Rydell, J. & Speakman, J. R. 1995. Evolution of nocturnality in bats: potential competitors and predators during their early history. *Biological Journal of the Linnean Society* 54, 183-191.
- Speakman, J. R. 1991. Why do insectivorous bats in Britain not fly in daylight more frequently? *Functional Ecology* 5, 518-525.
- Winter, Y., López, J. & von Helversen, O. 2003. Ultraviolet vision in a bat. *Nature* 425, 612-614.

Chapter 39:
Basal tyrant dinosaurs and my pet *Mirischia*

A few introductory comments for novices before I begin. Theropoda is the group name for the predatory dinosaurs (including birds), and Coelurosauria is a major theropod group that includes birds and all the bird-like theropods (including tyrannosauroids). Tyrannosauroidea includes the familiar giant tyrannosaurs like *Tyrannosaurus* of the Upper Cretaceous as well as an assortment of less familiar theropods, the oldest of which are from the Upper Jurassic.

So is it time to produce the definitive blog post on *Eotyrannus lengi*, the dinosaur I did my PhD on? Maybe. Actually, no. *Eotyrannus* was named by myself and colleagues in 2001, and in that initial paper we proposed that it was a basal tyrannosauroid, and one of the most basal members of the group (Hutt *et al.* 2001). Since then I've described the anatomy of *Eotyrannus* in full and tedious detail (the relevant thesis chapter is 118 pp and over 30,000 words long) and have come to know it well. I am now utterly convinced that it is a tyrannosauroid, and the results of my cladistic analysis (and those of others – see Holtz 2004) support this. Every dinosaur expert who knows anything about *Eotyrannus* agrees, by the way.

It turns out that the 2001 characterisation of *Eotyrannus* is horrendously wrong, as a new rigorous skeletal reconstruction (to be published soon) shows. The animal looked substantially different from the way I initially reconstructed it (see Naish 2001, Naish *et al.* 2001 and Holtz 2004), and in detail it's proved to be strikingly odd and unique in many, many features. I'll talk about these details some time soon, but not now. In fact I'm currently making arrangements to get the full monograph published in a high prestige journal, and with the co-authorship of a leading expert on tyrant dinosaurs I hope to produce an important work on tyrannosauroid phylogeny and morphology. More on this as and when it happens.

Since *Eotyrannus* was published a few very interesting things have been happening in the world of basal tyrannosauroids. I was aiming to discuss all of these here, but as usual I veered off at a tangent and have hardly scratched the surface. *Eotyrannus* seems to have been a mid-sized theropod. The type specimen is a juvenile individual that would have been 4.5 m long when complete, but fragmentary specimens from larger individuals indicate that adults were perhaps around 7 m long. But other basal tyrannosauroids are way smaller than this, mostly being less than 3 m long.

Fig. 115. *Dilong paradoxus* as reconstructed by Portia Sloan.

We begin with *Dilong paradoxus*, a basal tyrannosauroid known from excellent near-complete specimens from the Lower Cretaceous Yixian Formation of Liaoning Province, China (Xu *et al.* 2004). The Yixian Formation is the now famous unit that has produced all those little coelurosaurs with feathers and other integumentary structures preserved, and *Dilong* is no exception. It is preserved with simple quill-like integumentary structures that seem antecedent to the true complex feathers that evolved later.

Dilong has proved to be a sort of Rosetta stone for me, allowing several previously enigmatic Lower Cretaceous coelurosaurs to be reinterpreted as additional basal tyrannosauroids. Given that I haven't published the relevant details, I'd be silly if I gave the game away here, but to be honest the only people who really care about this already know the relevant details, and furthermore I am silly anyway. The news is that a controversial little coelurosaur from the Isle of Wight's Wessex Formation, *Calamosaurus foxi* (known only from two cervical vertebrae, one of them incomplete), is so similar to the cervical vertebrae of *Dilong* that I am confident that it too should be identified as a basal tyrannosauroid. This is mentioned in a large manuscript that came back from review some weeks ago and is currently undergoing revision, but the full story is to be revealed in a short paper that's been completed and reviewed but now awaits post-review revamping.

I previously had *Calamosaurus* down as a compsognathid (Naish *et al.* 2001). Compsognathids are, like tyrannosauroids, basal coelurosaurs, but they retained small body size throughout their history (so far as we know). They were also conservative in all being rather alike: morphologically unspecialized with relatively short forelimbs, rather long and gracile legs and feet, and a long tail. There's a lot more that could be said about them but this isn't the time. Here's the thing: the fact that *Calamosaurus* probably isn't one of them after all leads us to a key question. Namely, do all the *other* compsognathids really go together, or is Compsognathidae as currently perceived actually an artificial assemblage of distantly related (yet superficially similar) theropods?

Back in 2004 I and colleagues named the new Brazilian theropod *Mirischia asymmetrica.* That name means 'asymmetrical wonderful pelvis', and I think it's a rather good descriptive name. The only known specimen really is a 'wonderful pelvis' (though it also includes some of the hindlimb bones), as it is 3-D and fantastically well-preserved (Fig. 116), including even soft tissues like part of the gut and a probable post-pubic air sac (and to know the importance of that latter feature you'll have to wait). As you might guess, it's also asymmetrical, but I won't talk about that here. I also don't want to talk about the embarrassing fact that *Mirischia*'s generic name might imply that it was named after the fund-giving Mirisch Foundation, but (annoyingly) I only found this out after the publication of the *Mirischia* paper.

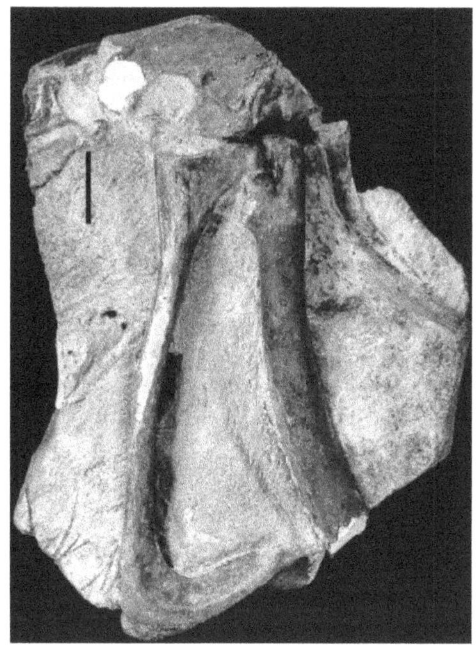

Fig. 116. The type specimen of *Mirischia asymmetrica:* a partial pelvis with some hindlimb bones. From Martill *et al.* (2000). The scale bar represents 20 mm.

Anyway, *Mirischia* is enough like *Compsognathus* – particularly in the detailed anatomy of its pubis – to convince me that it must belong together with *Compsognathus* in a little clade, and by definition this clade has to be called Compsognathidae. So in the 2004 paper, Naish *et al.* argued strenuously that (1) there is a little clade called Compsognathidae, the members of which can be united on the basis of shared derived characters, and (2) in addition to *Compsognathus*, this clade includes *Sinosauropteryx* and *Mirischia.* This published opinion is in part the result of a long series of to-ings and fro-ings between myself and my good friend Nick Longrich, for Nick has long been a vocal opponent of the idea that Compsognathidae really is monophyletic. Nick thinks that some so-called compsognathids are basal maniraptorans, and that others aren't even coelurosaurs, but to date he's only published an abstract on this (and to see why this is interesting and important you'll have to wait yet again. Think alvarezsaurids and stagodontids). Alas poor Longrich, surely he can't be right.

Here's where we bring *Dilong* back in. The pelvic anatomy of *Dilong* – an undoubted basal tyrannosauroid – is (like that of *Mirischia*) rather like that of *Compsognathus*. This is disturbing, as it all but destroys the reasons for thinking that *Mirischia* can *only* be a compsognathid. Might it actually be a basal tyrannosauroid? This has implications for another supposed compsognathid: the Isle of Wight taxon *Aristosuchus pusillus* which, again, I'd previously identified with confidence as a close relative of *Compsognathus* (Naish 2002, Naish *et al.* 2001, 2004). So we can now doubt that those 'compsognathids' known only from pelvic material really truly are compsognathids. Again, this is an idea that I'm mentioning in that large in-preparation manuscript.

In my thesis I tried to test all of this: I included all of the relevant taxa (except *Aristosuchus*, as I decided it wasn't complete enough to code) and all of the characters that have been used in this debate. And the result? Well, yes, there was a monophyletic Compsognathidae, but it consisted *only* of *Compsognathus* and one other taxon (and I'd rather not say which taxon that was right now). It wasn't *Mirischia*, as this came out as… a basal tyrannosauroid. Two characters helped pull *Mirischia* into Tyrannosauroidea. One of these was discussed in the Naish *et al.* (2004) paper but the other was previously overlooked. I'll not mention them here for fear of giving away all the secrets.

Furthermore, other supposed compsognathids did not group with Compsognathidae proper. Instead they were scattered about the base of Coelurosauria. So right now – while further work and further testing and further incorporation of data is needed – I am thinking that Nick was right, and that Compsognathidae in its old, inclusive sense is an artificial grouping.

People sometimes ask what sort of relevance stuff like this really has for our understanding of animals and their evolution. Well, it actually tells us an awful lot of stuff about patterns and trends. If so-called compsognathids – all of them relatively small, ecologically and morphologically generalized, long-limbed, long-tailed theropods that hunt small vertebrate prey – are not a clade but are actually scattered about the base of the coelurosaur family tree, this likely indicates that this ecotype was the ancestral one for coelurosaurs. We might already have thought that based on other lines of evidence, but this would help confirm it. There are indications that 'compsognathids' could make a living just about anywhere (for reasons that, again, I'll have to cover elsewhere), and if this is valid then again we have another really interesting discovery about evolution at the base of Coelurosauria.

Refs - -

- Holtz, T. R. 2004. Tyrannosauroidea. In Weishampel, D. B., Dodson, P. & Osmólska, H. (eds) *The Dinosauria, Second Edition*. University of California Press (Berkeley), pp. 111-136.
- Hutt, S., Naish, D., Martill, D. M., Barker, M. J. & Newbery, P. 2001. A preliminary account of a new tyrannosauroid theropod from the Wessex Formation (Early Cretaceous) of southern England. *Cretaceous Research* 22, 227-242.
- Martill, D. M., Frey, E., Sues, H.-D. & Cruickshank, A. R. I. 2000. Skeletal remains of a small theropod dinosaur with associated soft structures from the Lower Cretaceous Santana Formation of northeastern Brazil. *Canadian Journal of Earth Sciences* 37, 891-900.
- Naish, D. 2001. *Eotyrannus lengi*, a new coelurosaur from the Isle of Wight. *Dino Press* 5, 82-91.
- Naish, D. 2002. The historical taxonomy of the Lower Cretaceous theropods (Dinosauria) *Calamospondylus* and *Aristosuchus* from the Isle of Wight. *Proceedings of the Geologists' Association* 113, 153-163.
- Naish, D., Hutt, S. & Martill, D. M. 2001. Saurischian dinosaurs 2: Theropods. In Martill, D. M. & Naish, D. (eds) *Dinosaurs of the Isle of Wight*. The Palaeontological Asso-

ciation (London), pp. 242-309.

- Naish, D., Martill, D. M. & Frey, E. 2004. Ecology, systematics and biogeographical relationships of dinosaurs, including a new theropod, from the Santana Formation (? Albian, Early Cretaceous) of Brazil. *Historical Biology* 16, 57-70.
- Xu, X., Norell, M. A., Kuang, X., Wang, X., Zhao, Q. & Jia, C. 2004. Basal tyrannosauroids from China and evidence for protofeathers in tyrannosauroids. *Nature* 431, 680-684.

Chapter 40:

Dinosaurs come out to play

As a kid I always got the impression from textbooks that the only tetrapods (and thus only animals) that engage in play behaviour are (1) mammals and (2) a few really smart birds, like corvids and some parrots. Raptors are also known to engage in play behaviour, with it being relatively well documented that adults will drop feathers in front of their flying juveniles. The juveniles catch the feathers as if they're pretend prey.

But it would seem that play behaviour is not allowed to occur in lissamphibians, non-avian reptiles, or the majority of birds. They just don't do it, or at least no one has ever recorded them doing it. So why do mammals and oh-so-clever corvids and parrots, and predatory raptors, play, and why do other tetrapods not? Maybe so-called 'higher tetrapods' engage in play behaviour because full-blown endothermy allows this sort of superfluous, energy-wasting behaviour; maybe it's a result of enhanced encephalisation; or maybe it's only possible if extensive parental care allows juveniles enough behavioural 'security' to indulge in carefree behaviour.

Figure. 117. A Komodo dragon, photographed at ZSL London Zoo. Dragons can play too!

Well, here's the news. All of the above is crap. You might be surprised to hear that play behaviour is far from unique to mammals and a minority of birds, but has also been documented in turtles, lizards, crocodilians and even lissamphibians and fish (Bekoff 2000, Burghardt 2005). But because the reports discussing or mentioning play behaviour in these animals have been mostly anecdotal, and hence only mentioned as brief asides in larger behavioural studies or in brief one-page notes published in obscure journals, they have largely gone overlooked until recently.

Hold on: play behaviour in *reptiles, amphibians* and *fish*? Before looking at this further we need to sort out exactly what 'play' really is. How can it be defined? Of course this is something that ethologists have been arguing about for decades, and lengthy papers and virtually entire books (see Smith 1984 and Bekoff & Byers 1998) have been devoted to this topic alone. A rough working definition of play might be: a repeated behaviour, lacking an obvious function, initiated voluntarily when the animal is unstressed.

Most play behaviour – namely that observed in mammals and the more intelligent birds – is easily recognized by us because it resembles the sort of activities that we ourselves already recognize as playful. But this creates the obvious problem that play behaviour in other animals might be difficult to recognize because it is rather different from the sort of behaviours we 'expect' to represent play. Juvenile mammals tend to employ obvious honest signals when they're playing: we're all familiar with the 'play face' and bow-like action that canids (wild and domestic) use to initiate play, for example, and the play behaviour that they indulge in – chasing, play-biting, tussling and role-reversing – recalls human play behaviour.

However, if we employ the rough definition used above, behaviours reported widely among tetrapods can be seen in a new light. It turns out that several non-mammalian, non-avian vertebrates engage in repeated, apparently functionless behaviour that is initiated voluntarily in unstressed individuals. Sometimes this behaviour is directed toward inanimate objects (so-called manipulative play or object play).

Most of the key research in this area has been produced by Gordon M. Burghardt, and if you're interested in his research it's worth checking out his new book (Burghardt 2005). There's stuff here about apparent play behaviour in fish and – shock horror – even, outside of vertebrates, in cephalopods. I'm particularly interested in the play behaviour that's now been documented in captive trionychid and emydid turtles (Burghardt 1998, Burghardt *et al.* 1996, Kramer & Burghardt 1998).

Thinking about this reminded me of an activity indulged in by one of the Red-eared sliders *Trachemys scripta* we used to have in my UOP office. One of the terrapins used to regularly remove the plastic hose from the filter box in its tank, and then nudge the filter box around the tank. This was irritating as we (we = myself and Sarah Fielding) had to keep repositioning the box and reconnecting the hose. I honestly didn't think at the time that this behaviour 'meant' anything, but I'm wondering now if it was a form of play. Certainly those animals were bored with nothing to do in their little tank, so maybe they were in need of behavioural enrichment, and hence searching for objects to manipulate.

By introducing objects like wooden blocks and chains into enclosures, Burghardt and colleagues noted exactly this occurring in turtles, crocodilians and lizards. An Orinoco crocodile

Crocodylus intermedius rated particularly high in terms of its response to the objects, and appeared to exhibit both curiosity and playfulness toward them. There's also a published account of an American alligator *Alligator mississippiensis* exhibiting playful behaviour directed at dripping water (Lazell & Spitzer 1977), and there are also accounts of crocodilians possibly playing with carcasses, and apparently surfing in waves. I've seen a short sequence of film of two sibling Nile crocodiles *Crocodylus niloticus* tussling with one another in what looked like play behaviour.

The best data however comes from monitor lizards, and in fact from one individual monitor lizard in particular. Kraken is a well-studied female Komodo dragon *Varanus komodoensis* kept at the Smithsonian National Zoological Park in Washington, D. C. Developing a close bond with her keepers, it began to be noticed that she directed an unusual amount of curiosity toward shoe laces and to objects concealed in people's pockets (such as handkerchiefs and notebooks). Kraken would tug at or sever shoe laces (with her teeth), and would gently pull objects out of people's pockets. The keepers then began to introduce boxes, blankets, shoes and Frisbees into Kraken's enclosure, and many of Kraken's reactions would be interpreted as playful if witnessed in a mammal. Kraken has also been recorded to play tug-of-war with her keepers.

In a detailed, thorough study of Kraken's interactions with objects and her keepers, Burghardt *et al.* (2002) concluded that play-like behaviour in Komodo dragons definitely meets the formal criteria for play: "Kraken could discriminate between prey and non-prey and showed varying responses with different objects (i.e., ring and shoe). Large lizards, such as the Komodo dragon, might be revealed as investigative creatures, and further expressions of play-type behaviors should be confirmed and explored. These findings would imply that non-avian reptiles in general and large long-lived species in particular are capable of higher cognition and are much more complex than previously thought" (p. 116). It's interesting to note that probable play behaviour was reported in Komodo dragons as early as 1928, incidentally. Other people have now documented play behaviour in captive monitors.

Fig. 118. Dinosaurs come out to play.

So – if you'll excuse me here for bringing in some vertebrate palaeontology – did non-avian dinosaurs play? Several authors have speculated about this, but only in fictional essays: Stout & Service (1981) depicted baby tyrannosaurs chasing, wrestling and play-biting one another, and Bakker (1995) imagined dromaeosaurids and troodontids sliding down snowy slopes in a Cretaceous winter. Of course we don't know whether dinosaurs played, and we never will, but given how widespread play behaviour is in living reptiles, phylogenetic bracketing indicates that at least some extinct dinosaurs almost certainly would have engaged in this. So, artists, feel free to depict baby dromaeosaurs running around with feather or stick toys in their mouths.

Refs - -

- Bakker, R. T. 1995. *Raptor Red.* Bantam Press, London.
- Bekoff, M. 2000. The essential joys of play. *BBC Wildlife* 18 (8), 46-53.
- Burghardt, G. M. 1984. On the origins of play. In Smith, P. K. (ed). *Play in Animals and Humans.* Basil Blackwell, Oxford, pp. 5-41.
- Burghardt, G. M. 1998. The evolutionary origins of play revisited: lessons from turtles. In Bekoff, M. & Byers, J. A. (eds). *Animal Play: Evolutionary, Comparative, and Ecological Perspectives.* Cambridge University Press, Cambridge, pp. 1-26.
- Burghardt, G. M. 2005. *The Genesis of Animal Play: Testing the Limits.* MIT Press, Cambridge, MA.
- Burghardt, G. M., Chiszar, D., Murphy, J. B., Romano, J., Walsh, T. & Manrod, J. 2002. Behavioral complexity, behavioral development, and play. In Murphy, J. B., Ciofi, C., de La Panouse, C. & Walsh, T. (eds) *Komodo Dragons: Biology and Conservation.* Smithosonian Institution Press (Washington, DC), pp. 78-117.
- Burghardt, G. M., Ward, B. & Rosscoe, R. 1996. Problem of reptile play: environmental enrichment and play behavior in a captive Nile soft-shelled turtle, *Trionyx triunguis. Zoo Biology* 15, 223-238.
- Kramer, M. & Burghardt, G. M. 1998. Precocious courtship and play in emydid turtles, *Ethology* 104, 38-56.
- Lazell, J. D. & Spitzer, N. C. 1977. Apparent play behavior in an American alligator. *Copeia* 1977, 188-189.
- Smith, P. K. 1984. *Play in Animals and Humans.* Basil Blackwell, Oxford.
- Stout, W. & Service, W. 1981. *The Dinosaurs.* Bantam Books, New York.

Chapter 41:
Ichthyosaur wars and marvellous mixosaurs

When giving talks about ichthyosaurs – the 'fish lizards' of the Mesozoic – I always cover the basics: stuff such as, while they hung on until as late as the early part of the Late Cretaceous, they should best be regarded as animals of the Triassic and Early Jurassic as this is when their diversity was at its peak. Stuff such as the hyperphalangy and polydactyly that evolved in the limbs of some lineages, the well-known story of how soft-tissue-bearing specimens were first discovered, and all that data from Holzmaden (and other places) on ichthyosaur birth and babies. But there's a lot more stuff that doesn't get repeated so often.

Exploding whales, breech babies and toxic shock

On birth and babies, I contend that not all females 'preserved in the act of giving birth' really were giving birth when they died. Instead these individuals may have died while pregnant, with decomposition gases later pushing unborn babies out of the cloaca. Exactly this occurs in the dead bodies of beached whales today: pregnant females may have babies protruding from the birth canal, and males often have a distended penis that, similarly, has been extruded from the body cavity by gases building up inside. Of course this leads us on to the subject of exploding whales, but we won't go there for now. I have some nice anecdotes.

Some ichthyosaurs had breech births, as their babies are preserved protruding head-first. Here again we have an analogy with cetaceans. Baby whales and dolphins ordinarily emerge tail-first, and are thus only 'triggered' to take their first breath when the head emerges. But if the head emerges first, the baby drowns, and its little corpse is then lodged in the mother's birth canal. The mother then becomes slowly poisoned as the baby decomposes wedged inside her, and she dies of toxic shock. It's not nice, but it happens, and it's a reasonable (albeit untestable!) speculation to think that breech-birth ichthyosaur mothers sometimes died the same way too. If I remember correctly this idea was first proposed by Deeming *et al.* (1996), and I've a feeling that Naish (1997) picked up on it.

Marvellous mixosaurs

Some of the neatest new data on ichthyosaurs comes from newly appreciated taxonomic diversity. Mixosaurs are a fairly well studied and long-known group of basal Triassic ichthyosaurs, best known for little *Mixosaurus* (total length c. 1.5 m) named in 1887 for specimens from Middle Triassic Europe. Mixosaurs have always been depicted as rather dull and conventional. But it now seems that at least some of them were bizarre. Really really bizarre.

Middle Triassic Europe, North America and Spitsbergen was home to the mixosaur *Phalarodon,* named by John Campbell Merriam in 1910. At the back of its jaws are massive, rounded crushing teeth (properly known as tribodont teeth): proportionally huge, and in fact proportionally among the biggest of any ichthyosaur. The teeth at the jaw tips were slender and subconical, so *Phalarodon* seems to have been a generalist, perhaps picking up small soft-bodied prey with the rostral teeth, and crushing big hard-shelled prey with the tribodont teeth further back. Incidentally, a huge percentage of Triassic marine reptiles had tribodont crushing teeth like *Phalarodon*, and it's a good question as to why this was so common at the time, and so much rarer afterwards.

What also makes *Phalarodon* interesting is the presence of a proportionally large sagittal crest on the back of its head. Strongly compressed laterally and projecting dorsally from the skull roof to a height similar to that of the cranium itself, it must have had an important function, but we aren't too sure what that was. A site for muscle attachment is the most popular explanation.

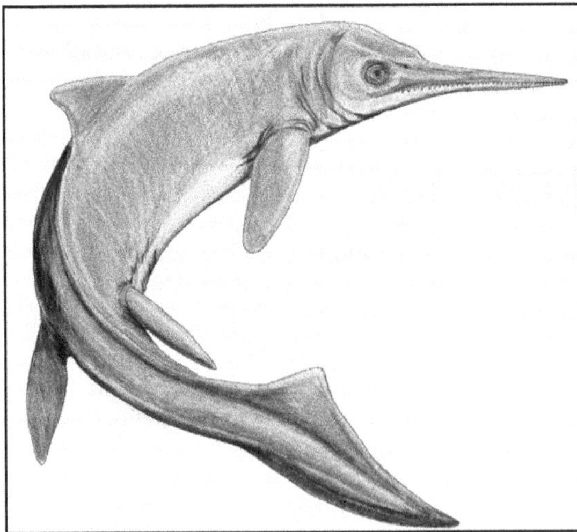

Fig. 119. The mixosaurid *Contectopalatus atavus*, as reconstructed by Dmitry Bogdanov.

Like *Mixosaurus*, *Phalarodon* wasn't particularly big, with *P. major* from Germany getting to perhaps 3.5 m. But the best is yet to come. The weirdest mixosaur – and, in my opinion, the weirdest ichthyosaur – is the freakish *Contectopalatus atavus*. Only known from the Middle Triassic of Germany, it was a giant compared to other mixosaurs, with some incomplete specimens indicating complete lengths of 5 m. Its skull was slender-jawed and, while its many sub conical teeth were blunt-tipped, it lacked the huge tribodont teeth of *Phalarodon.* It seems not to have gone around crushing molluscs or prey like that, therefore. It also has a sagittal crest, but it's even more prominent than that of *Phalarodon.* Sticking from the top of the skull like a piece of card, the sagittal crest seems to have been flanked by shallow concavities on the skull roof. Again, all of this may have been for muscle attachment, but nobody's really sure.

A big, mysterious and bizarre ichthyosaur, *Contectopalatus* was originally recognised as a new species in the 1850s, but not until 1998 did Michael Maisch and Andreas Matzke name it as a new genus (Maisch & Matzke 1998). For additional data on it, see Maisch & Matzke (2000a, b, 2001). Their reconstruction of its skull is shown in Fig. 120. It's at this point that I should note that not all ichthyosaur experts agree that *Phalarodon* and *Contectopalatus* are truly distinct from boring little *Mixosaurus*. Ryosuke Motani has strongly disagreed with this classification, and argues that all three forms should be synonymised (Motani 1999). Indeed Motani and Maisch & Matzke differ in their opinions on so many matters of ichthyosaur taxonomy and phylogeny that we talk of the 'Ichthyosaur wars', though it's not as if the workers involved would ever get physically aggressive with one another (I assume). Motani is a student of Chris McGowan, or 'god' as those in the ichthyosaur research community sometimes call him.

Whatever its taxonomic status, there's no denying that *Contectopalatus* was unusual and interesting. This begs the question as to why it's not better known: I have yet to see a single artistic restoration of it, for example (I wrote this before Fig. 119 had appeared online). Back when the BBC were still deciding which animals they were going to include in the *Sea Monsters* series (fronted by Nigel Marven) they screened in 2003, I (via Dave Martill, one of their technical consultants) strongly recommended use of *Contectopalatus*. But they didn't go with it. Shame. So there it sits, in the literature, unexploited and largely unknown.

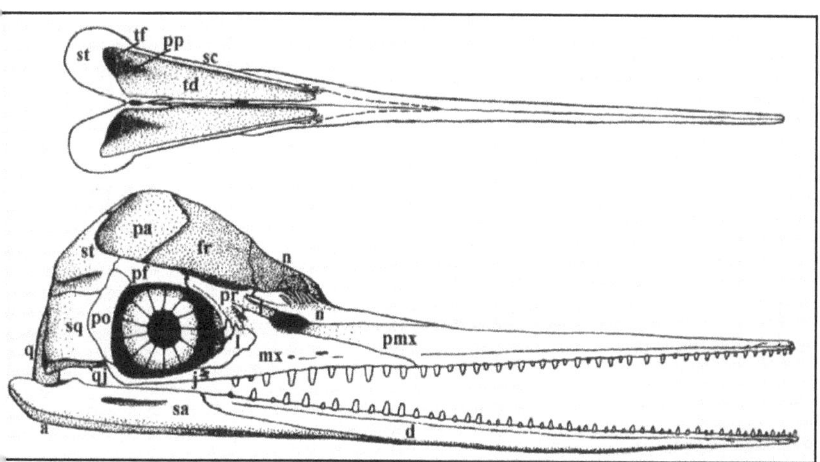

Fig. 120. The bizarre skull of *Contectopalatus,* as reconstructed by Maisch & Matzke (2000b).

At 5m in length, *Contectopalatus* is reasonable in size for a Triassic ichthyosaur, but it's not exceptional. The cymbospondylids and shastasaurs – located closer to Jurassic ichthyosaurs on the cladogram than the mixosaurs – grew to larger sizes and were also far more formidable: their stout, keeled teeth and robust jaws indicate that they were macropredators that perhaps filled the role that pliosaurs and mosasaurs did later on in the Mesozoic. And it's among shastasaurs that we find the biggest of all ichthyosaurs, and indeed the biggest of all marine reptiles. I was going to talk about them here, but now I can't. I was also going to talk about the swordfish that speared Alvin the DSRV and about *Excalibosaurus* and *Eurhinosaurus* and about so much else, but it will have to wait to another time.

Refs - -

- Deeming, D. S., Halstead, L. B., Manabe, M. & Unwin, D. M. 1995. An ichthyosaur embryo from the Lower Lias (Jurassic: Hettangian) of Somerset, England, with comments on the reproductive biology of ichthyosaurs. In Sarjeant, W. A. S. (ed) *Vertebrate Fossils and the Evolution of Scientific Concepts*. Gordon and Breach Publishers, pp. 463-482.
- Maisch, M. W. & Matzke, A. T. 1998. Observations on Triassic ichthyosaurs. Part III: A crested, predatory mixosaurid from the Middle Triassic of the Germanic Basin. *Neues Jahrbuch fur Geologie und Palaontologie, Abhandlungen* 209, 105-134.
- Maisch, M. W. & Matzke, A. T. 2000a. The Ichthyosauria. *Stuttgarter Beiträge zur Naturkunde Serie B (Geologie und Paläontologie)* 298, 1-159.
- Maisch, M. W. & Matzke, A. T. 2000b. The mixosaurid ichthyosaur *Contectopalatus* from the Middle Triassic of the German Basin. *Lethaia* 33, 71-74.
- Maisch, M. W. & Matzke, A. T. 2001. The cranial osteology of the Middle Triassic ichthyosaur *Contectopalatus* from Germany. *Palaeontology* 44, 1127-1156.
- Motani, R. 1999. The skull and taxonomy of *Mixosaurus* (Ichthyopterygia). *Journal of Paleontology* 73, 917-928.
- Naish, D. 1997. *Aspects of Ichthyosaur Evolution and Ecology With Comments on Cross-Taxon Convergence Seen Throughout Marine Tetrapods*. Research Project Report 1997/97, Department of Geology, University of Southampton, pp. 80.

Chapter 42:

'A miniature plesiosaur without flippers': surreal morphologies and surprising behaviours in sea snakes

Though I have no experience with them whatever (the shame), I have a great affection for sea snakes. They tie together several of my favourite areas: the discovery of new species and cryptic species diversity; radical convergence, evolutionary reversals, and a surprising phylogeny; unexpectedly complex behaviours; seriously bizarre morphology and novel trophic adaptations.

Fig. 121. Cartoons representing the three main sea snake groups. At top, *Emydocephalus*. *Hydrelaps* is at bottom right and a sea krait (*Hydrophis*) at bottom left.

So when I had to give an introduction to the evolution and diversity of marine reptiles at an educational conference held at Vaughan College, Leicester, in 2004 I decided, for reasons that made sense at the time, to cover sea snakes as well. Because the conference was otherwise devoted to Mesozoic reptiles this confused at least some people in the audience (well, it confused Richard Forrest anyway) as they then thought that sea snakes had evolved in the Mesozoic.

Well, actually, there *were* 'sea snakes' in the Mesozoic – that is, marine members of Serpentes – but they weren't close relatives of modern sea snakes, rather they were far more basal within the snake family tree (though how basal is the subject of contentious debate). Sea snakes in the modern sense are proteroglygous (= 'front-fanged') caenophidians that belong to Elapidae, the widespread and successful snake clade (consisting of about 300 species) that includes cobras, coral snakes, mambas and the Australasian tiger snakes, taipans, brown snakes, whip snakes and so on.

Given that the oldest fossil elapids – cobras, coral snakes and other forms from Germany, Spain and the Czech Republic (Ivanov 2002) – are from the Early Miocene, it's generally thought that sea snakes can't be older than this. In fact a supposed fossil sea snake, represented only by vertebrae, has been reported from the Middle Miocene of the former USSR. Rage (1987) noted that this record was questionable given that sea snake vertebrae "are not easily distinguished from those of the Colubridae and other Elapidae" (p. 66).

Conventional thinking has been that: if sea snakes descended from terrestrial elapids, we shouldn't then expect them to have a history extending beyond the Early Miocene. But for various reasons that I'm not going to discuss here, some workers are now suggesting that modern sea snakes may, after all, have evolved in the Cretaceous (see Rasmussen 2002).

Indeed the way sea snakes are related to other elapids is interesting, to say the least. Though sea snakes all possess a vertically flat, paddle-like tail (a feature not seen in any other snakes), most snake experts have agreed that the term 'sea snake' includes two quite different groups of aquatic elapids: the laticaudids, or sea kraits, and the hydrophiids (or hydrophids or hydro-pheids), or true sea snakes. In fact, even the paddle-like tail isn't really alike in laticaudids and hydrophiids, as in the former the neural spines aren't elongate as they are in hydrophiids (Rasmussen 1997). Several other features indicate that laticaudids are 'primitive' compared to the hydrophiids: they are generally better able to move on land, and they are mostly (but not entirely) oviparous (whereas hydrophiids are all viviparous).

While some herpetologists have regarded laticaudids as close relatives of sea snakes proper, McDowell (1969, 1987) argued that laticaudids were most closely related to Asiatic coral snakes (*Calliophis* and *Maticora*), American coral snakes (*Micrurus* and *Micruroides*) and *Parapistocalamus* (a poorly known snake of New Guinea and Bougainville Island: it's apparently a specialist eater of snail eggs), and this has since been quite widely supported by other snake workers (e.g., McCarthy 1986, Keogh 1998, Rasmussen 2002). Sea snakes therefore represent two separate invasions of the marine environment. However, the news is that the picture is even more complicated than this, as it now seems that hydrophiids themselves are not monophyletic.

You might be surprised to hear how big sea snakes get, and how odd some of them are in shape. Heuvelmans (1968) noted that members of the genera *Hydrophis* and *Microcephalophis* [nowadays included within *Hydrophis*] "have a long thin head and neck, while the abdomen is four or five times as thick as the neck, so that they look like a miniature plesiosaur, but without its flippers" (p. 38). He also noted that some *Hydrophis* species can be as much as 3 m long, and he cited William Dapier's 18th century observation of an individual "as big as a man's leg" (which, if applying to circumference, seems massive). I checked some of these details with Arne Rasmussen, a sea snake expert based at the School of Conservation, Copenhagen, who confirmed that a marked disparity between abdominal and neck circumference genuinely is present in some species. Is 3 m an accurate total length? Both Blue-banded sea snakes *H. cyanocinctus* and Yellow sea snakes *H. spiralis* have an authenticated maximum length of 2.75 m, so they're not far off from this. Most of the 50-odd species are between 1 and 2 m long, however.

Not only do some species grow rather large, some species form unbelievably large breeding aggregations, with hundreds, thousands and apparently millions of individuals sometimes grouping together to form immense slicks that can be literally kilometres long. In *The Trail That is Always New* (1932), Willoughby Lowe described a sea snake slick encountered between Sumatra and the Malayan Peninsula. It ran parallel to the ship he was travelling in for a duration of about 60 miles, and for part of its length was at least 3 m wide. Lowe was probably justified therefore in stating that "Along this line there must have been millions" (Bright 1989). To my knowledge, sightings of such super-aggregations remain anecdotal and they haven't been photographed. If you know otherwise please let me know!

Sea snakes also include what is often said to be the most venomous of all snakes, *H. belcheri*. Its venom is said to be several times more potent than that of the Fierce snake *Oxyuranus microlepidotes* – the most venomous land snake – but it's a very docile species that, even when handled roughly, rarely bites, and when it does bite it doesn't inject much venom. Like sea snakes in general it has only small fangs and it's not able to strike out of water. I couldn't find any neat statistics on *H. belcheri* venom (such as "one drop of its venom could kill the entire population of an average Texan trailer park"), but for comparison an exceptional 110 mg venom yield from a Fierce snake was reportedly enough to kill 250,000 mice (Carwardine 1995).

Many sea snake species seem to use estuarine habitats as breeding grounds and/or nurseries, but they aren't restricted to marine environments and may travel for tens of kilometres up rivers. Species in Cambodia, Thailand and the Philippines have been reported from lakes, and in fact two species are restricted to lakes: one of these is a sea krait that inhabits Rennell Island in the Solomons, and the other is *H. semperi* of Lake Taal on Luzon Island. What might be an additional freshwater species, *H. sibauensis*, was described by Rasmussen *et al.* (2001) for specimens collected 1000 km up-river in the River Sibau of Kalimantan, Indonesia. This is the furthest any sea snake has ever been recorded from the sea, but Rasmussen *et al.* (2001) were unable to determine whether the species was a true freshwater specialist as all known specimens were pregnant, and they might therefore have swam up-river in order to give birth. Even normally marine sea snakes species can switch between living in salt- or freshwater without problem, and have been kept in freshwater conditions without ill effects, so their physiology is pretty flexible.

Sea snakes are also neat in that they practice cutaneous respiration, being able to absorb up to 0% of their oxygen requiremens through their skin. That's right: cutaneous respiration isn't

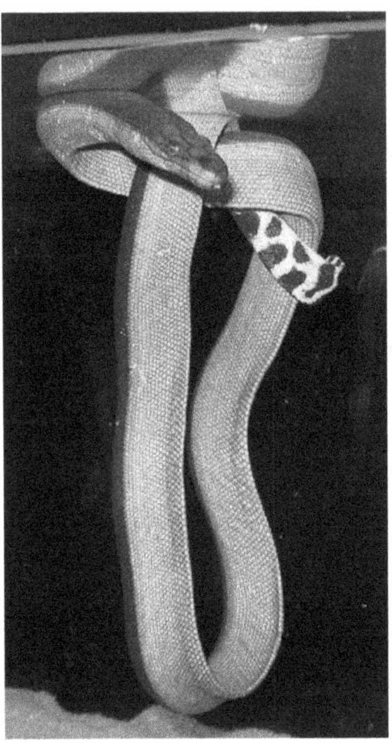

Fig. 122. Yellow-bellied sea snake *Pelamis platurus* photographed in captivity.

limited to lissamphibians among tetrapods. They can stay submerged for up to 3.5 hours and might only surface for 1 second (literally): in fact Yellow-bellied sea snakes *Pelamis platurus* (Fig. 122), the only truly pelagic sea snake, spends an *average* of 87% of its entire time submerged. Sea snakes are entirely absent from the Atlantic, but the pelagic habits of *Pelamis* have long led people to think that it might be able to use the freshwater Panama Canal to undergo a so-called Lessepian migration to the Atlantic. This hasn't happened however, nor have sea snakes colonised the Red Sea, let alone used the Suez Canal to get into the Mediterranean. Cool temperatures seem to have prevented *Pelamis* from extending its range around the Cape of Good Hope, as it seems to find sea temperatures of 11° C fatal.

Special soft-tissue valves allow sea snakes to close their nostrils while they're underwater and their lung is highly specialized. 'Lung', why not 'lungs'? In snakes generally the right lung is enlarged relative to the left lung (or, rather, the left lung has reduced in size), but in sea snakes the right lung is modified for air storage. Its posterior part (the so-called saccular lung) has unique thick, muscular walls* that allows stored air to be forced forward into the functional anterior part (termed the bronchial lung) and the anterior projection of the lung (the so-called tracheal lung) is large and extends well forwards relative to that of terrestrial snakes. The presence of the reduced left lung is variable in sea snakes: most species lack it, but it's present in some laticaudid individuals.

Sea kraits: radical intraspecific diversity, reproductive isolation, and site fidelity

We saw above that sea snakes aren't monophyletic, but that the two distinct groups (the laticaudids, or sea kraits, and the hydrophiids, or true sea snakes) instead represent two separate invasions of the marine environment. Furthermore, there are now indications that even hydrophiids aren't monophyletic.

Laticaudids are less specialised for marine life than hydrophiids, being better able to move on land, and they are mostly (but not entirely) oviparous (whereas hydrophiids are all viviparous). This might show that they've taken to marine life more recently than hydrophiids have, but then it's also possible that they're simply conservative. And don't go thinking that they're any

* Normally in snakes, the posterior lung walls are thin and sac-like, hence the name 'saccular lung'.

less interesting than hydrophiids. Thanks almost entirely to the field research of Richard Shine and Sohan Shetty of the University of Sydney, it's recently been discovered that laticaudids are actually rather complex, quite variable creatures. In fact they illustrate nicely Shine & Bonnet's (2000) point that snakes can be considered model organisms in terms of how much recent research has told us about evolution, ecology and behavioural diversity.

Six species of laticaudid are presently recognised. The Black-banded sea krait *L. laticaudata* (Linnaeus, 1758) is a large (1.1 m long), wide-ranging species. Populations differ in the patterning they have on the head, and while two subspecies have been distinguished on the basis of such differences, the variation actually appears to be clinal (McCarthy 1986). *L. crockeri* Slevin, 1934 is a denizen of Lake Te-Nggano on Rennell Island: it's one of the freshwater sea snakes I mentioned earlier. It's often melanistic. The Yellow-lipped sea krait *L. colubrina* (Schneider, 1799) (Fig. 123) is similar to *L. laticaudata* in distribution, and is highly variable across its range: in fact some workers have suggested that it should be split into as many as six subspecies. *L. semifasciata* (Reinwardt, in Schlegel, 1837) from the coasts of Japan, the Philippines, the Moluccas and the Lesser Sunda Islands is highly similar to *L. schistorhynchus* (Günther, 1874) from Niue, Tonga and Samoa, and some workers have considered them synonymous. Finally, there's the small (less than 1 m long) *L. frontalis* De Vis, 1905.

Among these species, *L. colubrina* and *L. frontalis* are virtually identical in morphology, differing only in size. The two are actually so similar that they have until recently been regarded as conspecific, despite detailed character analysis (McCarthy 1986), and not until Cogger *et al.*'s (1987) study was *L. frontalis* shown to be distinct. It had actually been first named by De Vis in 1905 and can be diagnosed on the basis of its small size, ventral scale counts and banding pattern.

Both *L. colubrina* and *L. frontalis* occur sympatrically around Vanuatu. Yet, despite their similarity, hybrids have never been reported, so how do these two remain reproductively isolated? It seems that males court females when triggered by the pheromones contained within the lipids that occur on female's bodies, and that they only court conspecifics (Shine *et al.* 2002). What's particularly interesting about this study from the point of view of science history is that, while sexual selection and isolating mechanisms among sympatric species have become heavily studied, virtually no research of this sort has been published on snakes.

What's also interesting is that the lipids concerned function in waterproofing. I assume then that they've been exapted to assist in pheromone distribution (that is, the lipids evolved for waterproofing first, and then became co-opted during evolution for pheromone distribution), but I can't determine this from Shine *et al.* (2002).

Laticaudids occur throughout the Indo-Australasian area, ranging from southern Japan and the Bay of Bengal in the north to Tonga and Somoa in the east, and to Tasmania and New Zealand in the south. Intriguingly, there are a few unconfirmed reports that the Yellow-lipped sea krait *L. colubrina* occurs on the Pacific coasts of Nicaragua, El Salvador and Mexico. The claim that this species has managed to cross the Pacific is pretty radical and really requires confirmation: as McCarthy (1986) stated "unfortunately these records are based on material that is no longer available for examination; the presence of *L. colubrina* in tropical America therefore requires substantiation" (p. 134).

It's usually implied in the literature that laticaudids are more flexible, in ecomorphological terms, than are hydrophiids in that they are not all marine: as mentioned above, the sometimes melanistic *L. crockeri* is restricted to the brackish Lake Te-Nggano (also called Lake Tegano) on Rennell Island (one of the Solomons). This isn't really accurate, however, given that some hydrophiids also occur in lakes. The Yellow-lipped sea krait *L. colubrina*, the most studied laticaudid, is reportedly quite agile on land (which fits nicely with amino acid data indicating that it is the most basal member of the group) while *L. crockeri* is rarely, if ever, seen on land. What does *L. colubrina* do when it's on land? It seems to rest here after feeding at sea (presumably using solar heat to aid digestion) and also sloughs its skin, lays it eggs, and mates on land.

L. colubrina individuals have been shown to exhibit strong site fidelity: even after feeding in deep water many km from land, they return to the same 'home' island (Shetty & Shine 2002a). 'Homing' behaviour has been reported among other snakes (usually in species that return year on year to the same overwintering site), but these are terrestrial species that have been shown to follow pheromone trails. How then do laticaudids find their way home? I don't think any-one knows and this is a fascinating topic for future research. Maybe they have a very good memory!

However, the terrestrial abilities of laticaudids don't just differ among species: they also differ among *individuals* of a species. Intraspecific variation in locomotor ability was studied in *L. colubrina* by Shine & Shetty (2001a) who found that males were substantially more agile on land than were females.

This sexual difference is probably due to the smaller size and more gracile proportions of males, but there might be intense selection for good terrestrial abilities in males because males do a lot of searching on land for female mates (interestingly, the good terrestrial abilities of males in this species is opposite to what's seen in sea turtles, where the males of some species may never ever leave the water, and are hence less able to perform terrestrial locomotion than are females).

Because they have to move on land a lot more

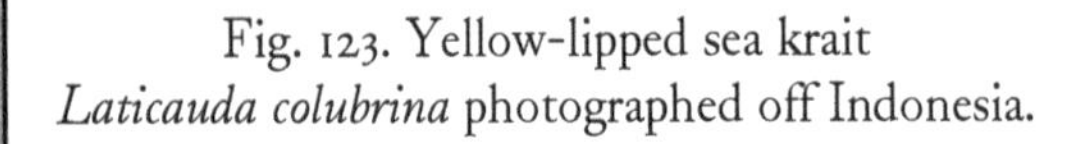

Fig. 123. Yellow-lipped sea krait *Laticauda colubrina* photographed off Indonesia.

than females do, male *L. colubrina* have relatively longer tails than females but there's a complex interplay between growth rate, survivability, and both aquatic and terrestrial agility in males (Shine & Shetty 2001b). In view of all this intraspecific variability, Shine & Shetty (2001a) noted how Yellow-lipped sea kraits might "offer exceptional opportunities to study phylogenetic shift in locomotor ability" because they "display considerable intraspecific and interspecific diversity in terms of the degree to which they use terrestrial vs. aquatic habits" (p. 338).

This reminds me of the intraspecific variation I discussed in Flying steamer-ducks *Tachyeres patachonica* (see Chapter 34): the species includes both flight-able and flightless individuals. Intraspecific variation in locomotor abilities among snakes is pretty special, however, elsewhere being documented only where pregnancy makes big females less agile than males, and in garter snakes where individuals that have only just emerged from hibernation are less agile than warmed-up, fully active indiduals (Shine *et al.* 2000).

Laticaudids are also intraspecifically variable in feeding behaviour. So far as we know all laticaudids are specialist predators of moray and conger eels, but it now seems that the sexes specialize on different types of eels. Again the species concerned is the Yellow-lipped sea krait, where the larger, more broad-headed females feed primarily on conger eels, while the smaller, more slim-headed males take the smaller moray eels (Shetty & Shine 2002b). Similar sexual variation in prey preference has been reported elsewhere in snakes in file snakes and American water snakes, so niche partitioning like this isn't unique to laticaudids.

So then... despite the fact that laticaudids aren't as specialised for marine life as hydrophiids are, they're clearly interesting. Is the amount of intraspecific varation we see in them unique to the group, or is it actually more widespread among sea snakes, but just undiscovered? That'll do on sea snakes for now. I really need to get back to my oviraptorosaurs.

Refs - -

- Bright, M. 1989. *There are Giants in the Sea*. Robson Books Ltd (London).
- Carwardine, M. 1995. *The Guinness Book of Animal Records*. Guinness Publishing (Enfield, Middlesex).
- Cogger, H. G., Heatwole, H., Ishikawa, Y., McCoy, M., Tamiya, N. & Teruuchi, T. 1987. The status and natural history of the Rennell Island sea krait, *Laticauda crockeri* (Serpentes: Laticaudidae). *Journal of Herpetology* 21, 255-266.
- Heuvelmans, B. 1969. *In the Wake of the Sea-Serpents*. Hill and Wang (New York).
- Ivanov, M. 2002. The oldest known Miocene snake fauna from Central Europe: Merkur-North locality, Czech Republic. *Acta Palaeontologica Polonica* 47, 513-534.
- Keogh, J. S. 1998. Molecular phylogeny of elapid snakes and a consideration of their biogeographic history. *Biological Journal of the Linnean Society* 63, 177-203.
- McCarthy, C. J. 1986. Relationships of the laticaudine sea snakes (Serpentes: Elapidae: Laticaudinae). *Bulletin of the British Museum of Natural History (Zoology)* 50, 127-161.

- McDowell, S. B. 1969. Notes on the Australian sea-snake *Ephalophis greyi* M. Smith (Serpentes: Elapidae, Hydrophiinae) and the origin and classification of sea-snakes. *Zoological Journal of the Linnean Society* 48, 333-349.
- McDowell, S. B. 1987. Systematics. In Seigel, R. A., Collins, J. T. & Novak, S. S. (eds) *Snakes: Ecology & Evolutionary Biology*. Macmillan (New York), pp. 3-49.
- Rage, J.-C. 1987. Fossil history. In Seigel, R. A., Collins, J. T. & Novak, S. S. (eds) *Snakes: Ecology & Evolutionary Biology*. Macmillan (New York), pp. 51-76.
- Rasmussen, A. R. 1997. Systematics of sea snakes: a critical review. *Symposium of the Zoological Society of London* 70, 15-30.
- Rasmussen, A. R. 2002. Phylogenetic analysis of the "true" aquatic elapid snakes Hydrophiinae (sensu Smith *et al.*, 1977) indicates two independent radiations into water. *Steenstrupia* 27, 47-63.
- Rasmussen, A. R., Auliya, M. & Bohme, W. 2001. A new species of the sea snake genus *Hydrophis* (Serpentes: Elapidae) from the river in West Kalimantan (Indonesia, Borneo). *Herpetology* 57, 23-32.
- Shetty, S. & Shine, R. 2002a. Philopatry and homing behaviour of sea snakes (*Laticauda colubrina*) from two adjacent islands in Fiji. *Conservation Biology* 16, 1422-1426.
- Shetty, S. & Shine, R. 2002b. Sexual divergence in diets and morphology in Fijian sea snakes *Laticauda colubrina* (Laticaudinae). *Austral Ecology* 27, 77-84.
- Shine, R. & Bonnet, X. 2000. Snakes, a new 'model organism' in ecological research? *Trends in Ecology & Evolution* 15, 221-222.
- Shine, R., Harlow, P. S., LeMaster, M. P., Moore, I. & Mason, R. T. 2000. The transvestite serpent: why do male garter snakes court (some) other males? *Animal Behaviour* 59, 349-359.
- Shine, R. & Shetty, S. 2001a. Moving in two worlds: aquatic and terrestrial locomotion in sea snakes (*Laticauda colubrina*, Laticaudidae). *Journal of Evolutionary Biology* 14, 338-346.
- Shine, R. & Shetty, S. 2001b. The influence of natural selection and sexual on the tails of sea-snakes (*Laticauda colubrina*). *Biology Journal of the Linnean Society* 74, 121-129.
- Shine, R., Reed, R. N., Shetty, S., Lemaster, M. & Mason, R. T. 2002. Reproductive isolating mechanisms between two sympatric sibling species of sea snakes. *Evolution* 56, 1655-1662.

Chapter 43:

The war on parasites: the pigeon's eye view, the oviraptorosaur's eye view

Don't take this the wrong way, but I love parasites, and if only there were more parasitic tetrapods I might get seriously, seriously interested in them. Not only is the biology and evolution of parasites really fascinating, the anti-parasite responses evolved by host species are too. And as we're going to see here, parasites night be so important to some tetrapods that their presence has exerted a significant evolution-ary pressure. In particular we're going to look at how birds have evolved to cope with certain ctoparasites. As we'll see, it may be that ectoparasites have had a significant effect on the volution of other tetrapod groups too.

eathers get dirty, damaged, stuck together and, perhaps most significantly, they harbour ecto-arasites, including feather lice, fleas, bugs, ticks and feather mites. Though there are bird pecies with specialized pedal claws that function in preening (namely herons, pratincoles and ightjars), birds rely on their bills when cleaning their feathers and removing ectoparasites. In act so important is the bill in keeping the feathers clean and relatively free of parasites that reening may – if the conclusions of some recent studies are to be accepted – be one of the ill's most important functions.

he over-riding factor controlling bill shape has, conventionally, been thought to be food type nd resource acquisition. Clearly this is still one of the most important, if not *the* most impor-ant factor controlling bill shape, otherwise we wouldn't have curlews, ibises, sword-billed ummingbirds, flamingos, crossbills, or all those oystercatcher polymorphs. But ornithologists ave lately started to notice that bill shape makes an awful lot of difference to parasite load. iven that parasites have been shown to have a major impact on fitness, and therefore on reeding success (Clayton 1990) and even on survival rate (Clayton *et al.* 1999), it follows nat anti-parasite adaptations might be really important.

One of the first studies to document this anti-parasite function was Clayton & Walther's (2001) on Peruvian Neotropical birds and their lice. Looking at species as diverse as owls, woodpeckers, barbets, jacamars, swifts, hummingbirds, pigeons, tyrant flycatchers, ovenbirds and swallows, they showed that those species with longer maxillary overhangs (viz, long edges to the upper mandibular tomia that overlap the edges of the lower mandibular tomia when the bill is closed) harboured less lice species than those species with short overhangs. Moyer *et al.* (2002) then noted that bill shape had an effect on parasite loads even within a single species: the Western scrub-jay *Aphelocoma californica* (Fig. 124).

Fig. 124. Western scrub-jay *Aphelocoma californica* of the *A. c. immanis* subspecies. This individual was photographed in Oregon.

Western scrub-jays are yet another example of resource polymorphism, an area discussed previously when relevant to oystercatchers (see Chapter 34). Those scrub-jays inhabiting oak woodland have hooked bills with long maxillary overhangs while those of pinyon-juniper woodlands have pointed bills with short overhangs: the oak woodland birds eat acorns while the pinyon-juniper woodland birds extract seeds from pine cones. Because they use their jaw tips as forceps to get the seeds out, the pinyon-juniper woodland birds seem to have secondarily reduced their overhangs. These two different bill shapes appear to correlate directly with louse control, as pinyon-juniper woodland Western scrub-jays have significantly more lice than the oak woodland Western scrub-jays, and this is despite the fact that oak woodland birds are physically larger and inhabitants of a more humid environment than pinyon-juniper woodland birds (Moyer *et al.* 2002). Pretty compelling stuff.

Dispatching certain ectoparasites – notably fleas and feather lice – isn't easy because the tough, flattened bodies of these arthropods are really good at resisting pressure. Simply grabbing the parasite and biting on it (thereby exerting vertical force onto the animal) isn't good enough, and Clayton & Walther (2001) proposed that the birds have to generate a shearing force in order to kill a captured parasite. Keep in mind that, once captured in the bill, the parasites do actually have to be killed, as if they're dropped they simply jump or climb straight back onto the host.

To test the idea that maxillary overhangs might function in parasite control, Clayton *et al.* (2005) trimmed the bills of juvenile Rock doves *Columba livia* (sorry, I can't bring myself to call them Rock pigeons [their new 'official' name]). This only involves removing 1-2 mm of the tomium by the way (Fig. 125) – it isn't anything like the brutal de-beaking indulged in by the factory chicken industry. Clayton *et al.* (2005) found that the trimming had no significant effect on the pigeon's feeding efficiency, so maxillary overhangs apparently do not exist for reasons related to feeding. But trimming did have a major impact on parasite load: trimmed birds were unable to keep their parasite loads down and exhibited a *significant* increase in feather damage relative to untrimmed birds. Trimmed birds that were allowed to regrow their overhangs "caused an immediate reduction in lice" (p. 815).

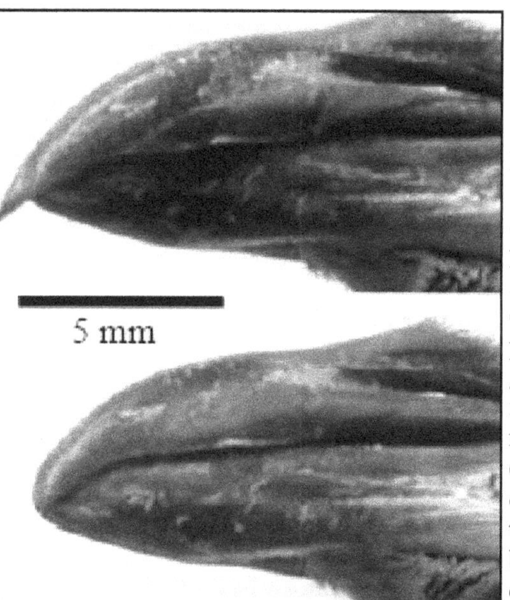

Fig. 125. Rock pigeon with (at top) a normal bill and (below) a bill where the maxillary overhangs have been trimmed.

Does the overhang work by allowing shearing of captured parasites? Using both high-speed video and data from strain gauge apparatus, it seems that pigeons move the lower jaw and upper maxillary overhang in concert, with the lower jaw exerting compressive strain against the overhang and generating a shearing force. This happens incredibly quickly, with the lower jaw being moved forward up to 31 times per second (Clayton *et al.* 2005, p. 815). The physical damage observed on lice killed by untrimmed pigeons was consistent with death by shearing: decapitation, lacerations of the exoskeleton and missing legs.

So the case, as demonstrated across a diverse range of avian taxa, looks pretty good. Maxillary overhangs really are important in parasite control, and the adaptive radiation of beak morphology should be re-assessed with both feeding *and* preening in mind.

But, like any interesting discovery, this now raises several new questions. Not all birds have maxillary overhangs: as Clayton *et al.* noted, many birds with specialized bills (including oystercatchers, darters, herons, woodpeckers, hummingbirds and scythebills) lack overhangs altogether, yet we know that these species have ectoparasites. We saw earlier how some birds have evolved pedal claws that probably function in preening, but given that these are also absent in some of the groups that lack overhangs, other defensive adaptations must be present. Some passerines are now known to be toxic (see Chapter 34), and it's been suggested that these toxins might function in parasite control (Mouritsen & Madsen 1994). In fact it's worth wondering if toxins are actually more widespread, and if they might be present in species that lack morphological structures that function in parasite control. Toxic oystercatchers? Well, maybe not, as birds can also use sunning, dust-bathing and other behaviours to control ectoparasites.

Obviously, Clayton *et al.* (2005) only considered what implications their study might have for living birds. But as a palaeontologist I'm going to do the logical thing and wonder what this might mean for *fossil* feathered taxa.

An oviraptorosaur's eye view

Fossil birds belonging to the same groups as extant species, surely, used their bills in the same manner as extant species, so dodos, teratorns and presbyornithids almost certainly found their bills to be as essential for preening as do modern pigeons, raptors and ducks. But of course we know that feathers weren't unique to 'modern-type' birds: they were also present in the basal birds of the Mesozoic (going all the way back to the archaeopterygids, and including a diverse aviary of yandangornithids, confuciusornithids, enantiornitheans, hesperornitheans and others) AND they were also present in non-avian maniraptoran theropods. We know that true feathers were present in oviraptorosaurs, microraptorians (which might be part of Dromaeosauridae: it depends on the phylogeny) and almost certainly troodontids (*Jinfengopteryx*, a luxuriantly feathered little theropod described in 2005 as an archaeopterygid, is almost certainly a troodontid). Furthermore, probable 'proto-feathers' (rather simple quill-like integumentary structures, almost certainly the morphological ancestors of true, complex feathers) were present in compsognathids, basal tyrannosauroids and alvarezsaurids.

We also know that ectoparasites were infesting feathers by the Cretaceous at least. How do we know this? Martill & Davis (1998, 2001) described an isolated feather from the Lower Cretaceous Crato Formation of Brazil that is covered in more than 240 hollow spheres that are almost certainly feather mite eggs. We also know that fleas were present in the Lower Cretaceous as there are two particularly good ones known from Australia (Riek 1970), and we also know of possible fleas and odd long-legged possibly parasitic insects from the Lower Cretaceous of the former USSR (Ponomarenko 1976). Terrestrial birds whose plumage is superficially similar to that of fuzzy small theropods are notorious for harbouring ectoparasites, with kiwis in particular being reported to crawl with numerous fleas, ticks, feather mites and lice (Kleinpaste 1991). So, I would be confident that Mesozoic birds, and fuzzy and feathered non-avian theropods, had to contend with ectoparasites. What then did they do about parasite control?

Unfortunately we don't know enough about the rhamphothecae of Mesozoic birds and bird-like maniraptorans to determine whether or not they had a maxillary overhang: the preservation simply isn't good enough. But maybe some of these animals didn't need a maxillary overhang given that many of them had teeth. Indeed several Mesozoic maniraptorans possess just a few teeth at the jaw tips, or even just in the upper jaw tips.

Fig. 126. The basal oviraptorosaur *Caudipteryx* as reconstructed by Matt Martyniuk.

Take the feathered turkey-sized short-skulled basal oviraptorosaurs *Protarchaeopteryx** and *Caudipteryx* (if you've heard that these animals aren't oviraptorosaurs but actually flightless birds, ignore it: it's a theory based on wishful thinking and misinterpretation of morphological evidence). In *Protarchaeopteryx*, teeth are restricted to the premaxillae and anterior parts of the maxillae and dentaries, with the premaxillary teeth being a few times taller than the others (Ji *et al.* 1998). In *Caudipteryx*, four procumbent teeth are present in each premaxilla, but the rest of the skull is edentulous. *Incisivosaurus* – closely related to, and possibly congeneric with, *Protarchaeopteryx* – has a reduced compliment of teeth, all of which are restricted to the anterior parts of the jaws, and two enlarged, bunny-like incisiform teeth project from each premaxilla. More derived oviraptorosaurs were toothless, but the bony premaxillary margins of their upper jaw were serrated, which raises the possibility that the tomium was serrated too. Could these serrations have been used in ectoparasite control?

Among other non-avian feathered maniraptorans, it's worth noting that microraptorians also exhibit an unusual premaxillary dentition. In *Sinornithosaurus*, a diastema separates the premaxillary teeth from the maxillary teeth, and the premaxillary teeth appear notably shorter than the maxillary ones (Xu & Wu 2001). While proportionally small premaxillary teeth are seen elsewhere in theropods (e.g., in tyrannosauroids), the combination of reduced dentition and diastema isn't, and we know without question that microraptorians had complex, vaned feathers on their limbs and tails. It's at least suggestive that the premaxillary teeth were used for preening.

Having mentioned tyrannosauroids, I might also note that basal forms combine proportionally

* Not a typo! I've lost count of how many times I've seen this name 'corrected' (to '*Protoarchaeopteryx*') by well-meaning editors.

small premaxillary teeth with quill-like integumentary structures that would have needed preening (or is grooming the correct term here?). Could those little premaxillary teeth have been specialized for ectoparasite control? I know this is grotesque speculation of the worst kind, but read on.

Moving now to Mesozoic birds, given that there was a trend in some lineages toward reduction and loss of teeth, it follows that members of these lineages exhibit reduced numbers of teeth relative to archaeopterygids and non-avian theropods. It seems that these birds lost the teeth from the back of the jaws first, and kept their premaxillary and dentary-tip teeth the longest. Even in forms that don't have a reduced dentition however, we see slight heterodonty, and thus some suggestion that the anterior-most teeth were being used for something special. In archaeopterygids for example, the premaxillary teeth are more peg-like and more procumbent than are the other teeth. *Aberratiodontus* – an odd relative of *Yanornis* from the Chinese Jiufotang Formation – has teeth lining both its upper and lower jaws, but is reported to have rather small teeth at the jaw tips (Gong *et al.* 2004).

When we start looking at some of the more unusual Mesozoic birds groups, we see marked specialisation of the rostral-most dentition. Bizarre long-tailed, robust-jawed *Jeholornis* (almost certainly synonymous with *Shenzhouraptor*, and perhaps with *Jixiangornis* too), known from stomach contents to have eaten seeds at least occasionally, has just three very small teeth at each lower jaw tip (Zhou & Zhang 2002): the upper jaw was edentulous.

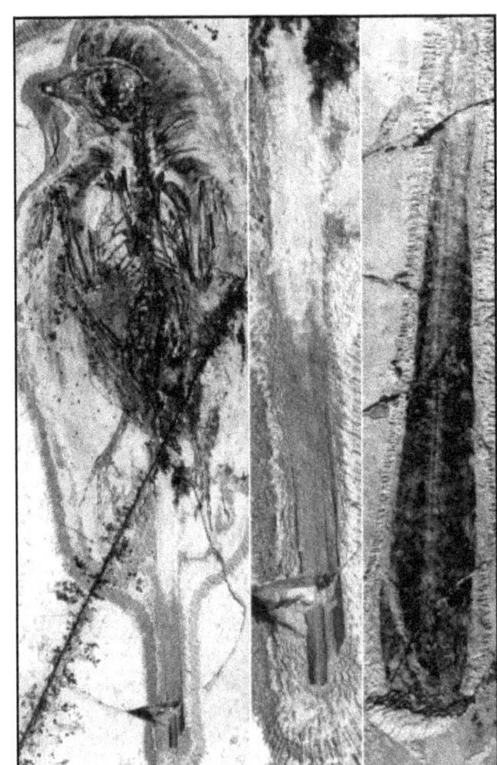

The unusual long-armed *Sapeornis*, also from the Jiufotang Formation, had a rather short, *Caudipteryx*-like skull, and short, conical, unserrated, procumbent teeth projected from its premaxillae (Zhou & Zhang 2003). Its dentaries were toothless (and its maxillae probably were too). The somewhat similar *Omnivoropteryx*, also from the Jiufotang Formation, was also short-skulled, and also has just a few procumbent teeth restricted to the premaxillae (Czerkas & Ji 2002).

Most (but not all) enantiornitheans were

Fig. 127. The Yixian Formation enantiornithean *Protopteryx,* famous for its strap-like tail feathers (shown in detail to the right of the specimen, but with one of the tail feathers of the confuciusornithid *Confuciusornis* shown at far right). From Zhang & Zhou (2000).

toothed, and ancestrally they had teeth lining their upper and lower jaws as archaeopterygids did. But in the Yixian Formation enantiornithean *Protopteryx* there are just two conical, unserrated teeth in the premaxillae and two subtriangular teeth at the dentary tips (Zhang & Zhou 2000). It doesn't seem that having a total of four teeth is a tremendously useful thing if you need those teeth to procure or dismember your food, and it's intriguing that *Protopteryx* sports highly elongate, strap-like rectrices that would (presumably) have needed careful preening (Fig. 127). The euenantiornithean *Eoenantiornis* has four subconical teeth in each premaxilla while there were probably six or seven teeth in each dentary, the rostral-most two of which were larger than the others (Zhou *et al.* 2005). Long-skulled *Longirostravis* has ten small, conical teeth restricted entirely to its slim jaw tips (Hou *et al.* 2004) and short, conical teeth are similarly only at the jaw tips in another long-skulled enantiornithean, *Longipteryx* (Zhang *et al.* 2001). And there are yet other examples of this sort of thing.

So far as I can see from all these unusual patterns of dentition, there are three possible explanations:- (1) as dental reduction occured, a gradual step-wise loss of teeth simply meant that premaxillary and/or dentary-tip teeth were the last to go; (2) premaxillary and/or dentary-tip teeth were retained in specialized taxa that used those teeth to procure or dismember whatever it was that they were eating; (3) premaxillary and/or dentary-tip teeth were retained – even when not essential to foraging or feeding – as they were used in ectoparasite control.

While it's nice to speculate – and so far that's all I've done here – how might we *test* the idea that these Mesozoic taxa were using their unusual rostral teeth to preen with? Herein lies the rub, as I can't think of a reliable test. So far as we know, feathers aren't abrasive enough to leave any sort of distinctive microwear on teeth, or even on rhamphothecae, so there isn't going to be any sort of tooth wear that can be correlated with preening. It's possible that there might be some sort of correlation between tooth spacing and feather morphology, but I find this unlikely.

Conversely, we *can* test the idea that teeth were used in feeding, as feeding does leave visible sorts of micro- or macrowear. Earlier I mentioned the bunny-like teeth of *Incisivosaurus*, and because its incisiform teeth do exhibit wear facets, they were almost certainly used in feeding. This confirms 'explanation 2' given above, and therefore indicates that 'explanation 3' didn't apply in this case. But the two 'explanations' aren't mutually exclusive, as the teeth could still have been important in ectoparasite control.

Those short premaxillary teeth present in tyrannosauroids have conventionally been regarded as having a primary role in feeding, and it might be easy to confirm this by looking for micro- or macrowear. And yes, I consider it *highly* speculative to wonder if those teeth might have functioned in grooming/preening, but I couldn't resist mentioning it (I want to discuss here the SEM data I have on the premaxillary teeth of the basal tyrannosauroid *Eotyrannus*, but that would require adding too many extra words).

Finally, if specialised teeth could be shown to have no important function in foraging or feeding behaviour it might then be logical to infer that preening was their primary function - - but, how on earth would you show that they had 'no function' in foraging or feeding behaviour?

This just isn't possible in Mesozoic animals when so little is known of their ecology. An analogy does come to mind: it's been shown that the unusual dentary teeth of Impala *Aepyceros melampus* have a morphology specialised for a primary function in grooming. If impalas were extinct I suppose it's possible that people might have worked this out, but how would you verify it? Does anyone have any better ideas?

I'm far from the first person to look at Mesozoic feathered theropods this way – many other people have mentioned these ideas before, and artists have even illustrated ectoparasite control in dinosaurs. In *Dinosaurs of the Air* Greg Paul illustrated a *Sinosauropteryx* scratching in order to remove ectoparasites, and the cover of *The Dinosauria, Second Edition* (the current industry-standard volume on dinosaurs) features a *Sinosauropteryx* (this time by Mark Hallett) nibbling at its proto-feathers, again presumably as a form of ectoparasite control.

But I don't think anyone's really married data on Mesozoic birds and other theropods with the new work of Dale Clayton and colleagues on ectoparasite control in extant birds. Maybe this idea will bear proverbial fruit down the line, but for now this is where my contribution ends.

Finally, here's another spin on this subject. Theropods weren't the only Mesozoic tetrapods with a furry coat of integumentary fibres: we also know that pterosaurs were fuzzy too. So did they also have to contend with ectoparasites? I'll say no more on this topic, but perhaps it can be elaborated on at another time.

Refs - -

- Clayton, D. H. 1990. Mate choice in experimentally parasitized rock doves: lousy males lose. *American Zoologist* 30, 251-262.
- Clayton, D. H., Lee, P. L. M., Tompkins, D. M. & Brodie, E. D. 1999. Reciprocal natural selection on host-parasite phenotypes. *American Naturalist* 154, 261-270.
- Clayton, D. H., Moyer, B. R., Bush, S. E., Jones, T. G., Gardiner, D. W., Rhodes, B. B,. & Goller, F. 2005. Adaptive significance of avian beak morphology for ectoparasite control. *Proceedings of the Royal Society London B* 272, 811-817.
- Clayton, D. H. & Walther, B. A. 2001. Influence of host ecology and morphology on the diversity of Neotropical bird lice. *Oikos* 94, 455-467.
- Czerkas, S. A. & Ji, Q. 2002. A preliminary report on an omnivorous volant bird from northeast China. In Czerkas, S. J. (ed) *Feathered Dinosaurs and the Origin of Flight* The Dinosaur Museum (Blanding, Utah), pp. 127-135.
- Gong, E., Hou, L. & Wang, L. 2004. Enantiornithine bird with diapsidian skull and its dental development in the Early Cretaceous in Liaoning, China. *Acta Geologica Sinica* 78, 1-7.
- Hou, L., Chiappe, L. M., Zhang, F. & Chuong, C.-M. 2004. New Early Cretaceous fossil from China documents a novel trophic specialization for Mesozoic birds. *Naturwissenschaften* 91, 22-25.
- Ji, Q., Currie, P. J., Norell, M. A. & Ji, S. 1998. Two feathered dinosaurs from northeastern China. *Nature* 393, 753-761.

- Kleinpaste, R. 1991. Kiwis in a pine forest habitat. In Fuller, E. (ed) *Kiwis: A Monograph of the Family Apterygidae.* Swan Hill Press (Shrewsbury), pp. 97-138.
- Martill, D. M. & Davis, P. G. 1998. Did dinosaurs come up to scratch? *Nature* 396, 528-529.
- Martill, D. M. & Davis, P. G. 2001. A feather with possible ectoparasite eggs from the Crato Formation (Lower Cretaceous, Aptian) of Brazil. *Neues Jahrbuch fur Geologie und Palaontologie, Abhandlungen* 219, 241-259.
- Moyer, B. R., Peterson, A. T. & Clayton, D. H. 2002. Influence of bill shape on ectoparasite load in Western scrub-jays. *Condor* 104, 675-678.
- Mouritsen, K. N. & Madsen, J. 1994. Toxic birds: defence against parasites? *Oikos* 69, 357-358.
- Ponomarenko, A. G. 1976. A new insect from the Cretaceous of Transbaikalia, a possible parasite of pterosaurians. *Paleontology Journal* 1976 (3), 339-43.
- Riek, E. F. 1970. Lower Cretaceous fleas. *Nature* 227, 746-747.
- Xu, X. & Wu, X.-C. 2001. Cranial morphology of *Sinornithosaurus millenii* Xu *et al.* 1999 (Dinosauria: Theropoda: Dromaeosauridae) from the Yixian Formation of Liaoning, China. *Canadian Journal of Earth Sciences* 38, 1739-1752.
- Zhang, F. & Zhou, Z. 2000. A primitive enantiornithine bird and the origin of feathers. *Science* 290, 1955-1959.
- Zhang, F., Zhou, Z., Hou, L. & Gu, G. 2001. Early diversification of birds: evidence from a new opposite bird. *Chinese Science Bulletin* 46, 945-949.
- Zhou, Z., Chiappe, L. M. & Zhang, F. 2005. Anatomy of the Early Cretaceous bird *Eoenantiornis buhleri* (Aves: Enantiornithes) from China. *Canadian Journal of Earth Sciences* 42, 1331-1338.
- Zhou, Z. & Zhang, F. 2002. A long-tailed, seed-eating bird from the Early Cretaceous of China. *Nature* 418, 405-409.
- Zhou, Z. & Zhang, F. 2003. Anatomy of the primitive bird *Sapeornis chaoyangensis* from the Early Cretaceous of Liaoning, China. *Journal of Paleontology* 40, 731-747.

Chapter 44:
'Angloposeidon', the unreported story

If *Mirischia* the Brazilian theropod is one of my little pets (see Chapter 39), the enormous sauropod represented by the cervical vertebra MIWG.7306 – affectionately (and unofficially) known to some of us as 'Angloposeidon' – is one of the biggest (Fig. 128 depicts my drawings of the specimen). I don't specialise on sauropods, but I sometimes collaborate on technical work on these awesome beasts. Mike P. Taylor and I co-authored a paper on the phylogenetic systematics of diplodocoids last year (Taylor & Naish 2005) and we have a few other sauropody projects in the pipeline, and Matt Wedel and I are going to be publishing sauropod stuff some time in the future.

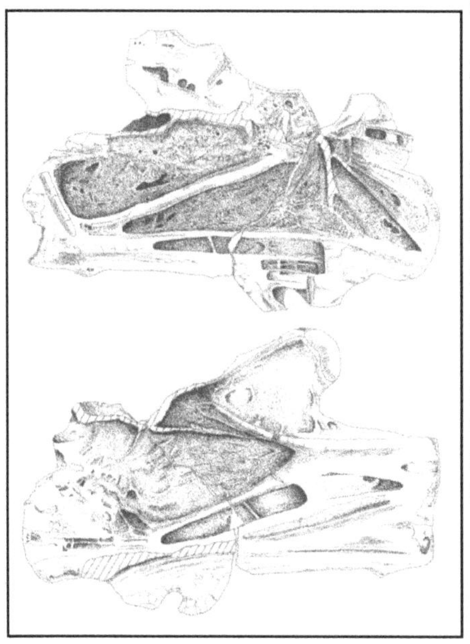

2000 was a really busy year for me, in part because of the huge Dorling Kindersley encyclopaedia I worked on, the *Dinosaurs of the Isle of Wight* book, my work on the preliminary description of *Eotyrannus* (Hutt *et al.* 2001), and the completion of my M.Phil. thesis. And a great deal of time that year was spent, with Dave Martill, running around the Isle of Wight looking at specimens we were planning to include in *Dinosaurs of the Isle of Wight*. On several occasions we stumbled over unreported specimens that we 'discovered' in collections, but in other cases we made specific trips to look at, and photograph, specimens that we'd seen discussed or described in the literature.

Fig. 128. The author's poor attempts at drawing both sides of MIWG.7306.

Among the latter was an allegedly enormous sauropod vertebra, mentioned on the last page of a booklet entitled '*The discovery of the island's largest dinosaur*' and produced by Dinosaur Farm Museum. Written anonymously (but actually penned, I think, by Steve Hutt), the booklet discusses the discovery and excavation of the Barnes High brachiosaurid, a specimen discovered by Steve in 1992 and currently on display at Dinosaur Isle Visitor Centre, Sandown. It's still owned by Dinosaur Farm Museum: a tourist attraction that lacks official museum credentials, the frustrating consequence being that the Barnes High brachosaurid is still unavailable for technical study. This is maddening as the specimen is easily the most complete European brachiosaurid, and in fact one of the best European sauropods (only little *Europasaurus* from Germany being better represented). The booklet (which includes no publication data at all and can only be cited as 'Anon. undated'!) states on p. 7....

> Last year an enthusiastic young collector, Gavin Leng, found and gave to the Sandown Museum a single giant neck vertebra which is 920mm (2'6") long. This was from an adult brachiosaurid which would have had a complete neck 8.8-11 metres (24-30 feet) long and a total length of 26 – 29.6 metres (70 – 80 feet)!

Highly intrigued, in the summer of 2000 I asked around. It turned out that, while the specimen was quite obviously a sauropod cervical vertebra, it had been discovered encased within a hard sideritic matrix, and literally years of careful preparation had been required to prep it out. It had been discovered, in 1992, on the foreshore at Sudmoor Point on the south-west coast of the Isle of Wight. Leng, its discoverer, is well known in the world of Isle of Wight palaeontology for finding a particularly good valdosaur specimen *and* for discovering *Eotyrannus*, so he's a pretty significant guy.

He had handed it over to what was then known as the Museum of Isle of Wight Geology, and from here it had been the pet project of David Cooper, a former pathologist and now amateur palaeontologist, who was working on the specimen at his house. I already knew David from many previous meetings and discussions we'd had about Isle of Wight dinosaurs, and when Dave (Martill) and I visited him during the summer of 2000 the specimen was looking pretty good, with clean bone surfaces almost entirely free of matrix.

We carefully carried it out on to David's back lawn and photographed it there on the grass, and one of the resulting images was published as Plate 14B in *Dinosaurs of the Isle of Wight*. David had not just been preparing the specimen, he had also been trying to interpret its morphology and affinities, and we decided then and there that we would collaborate on getting a description into print. It was easily the biggest British dinosaur vertebra I'd ever seen, and I knew that it was about on par with the similarly enormous cervical vertebrae of *Brachiosaurus*. In its degree of elongation it appeared similar too. I was later to learn that it had an accession number – MIWG.7306 – and I'll use that from hereon.

Early in 2000 I began corresponding with Mathew Wedel, then of the Sam Noble Oklahoma Museum of Natural History (but today of the University of California Museum of Paleontology). In March 2000 Matt published (with co-authors Richard Cifelli and R. Kent Sanders) his preliminary description of a giant brachiosaurid from the Antlers Formation of Oklahoma, named *Sauroposeidon proteles* (Wedel *et al.* 2000a). *Sauroposeidon* is known only from its immense cervical vertebrae, two of which are depicted in Fig. 129. A fuller description, with tons of new data, was published later in the year in *Acta Palaeontologica Polonica* (Wedel *et al.* 2000b).

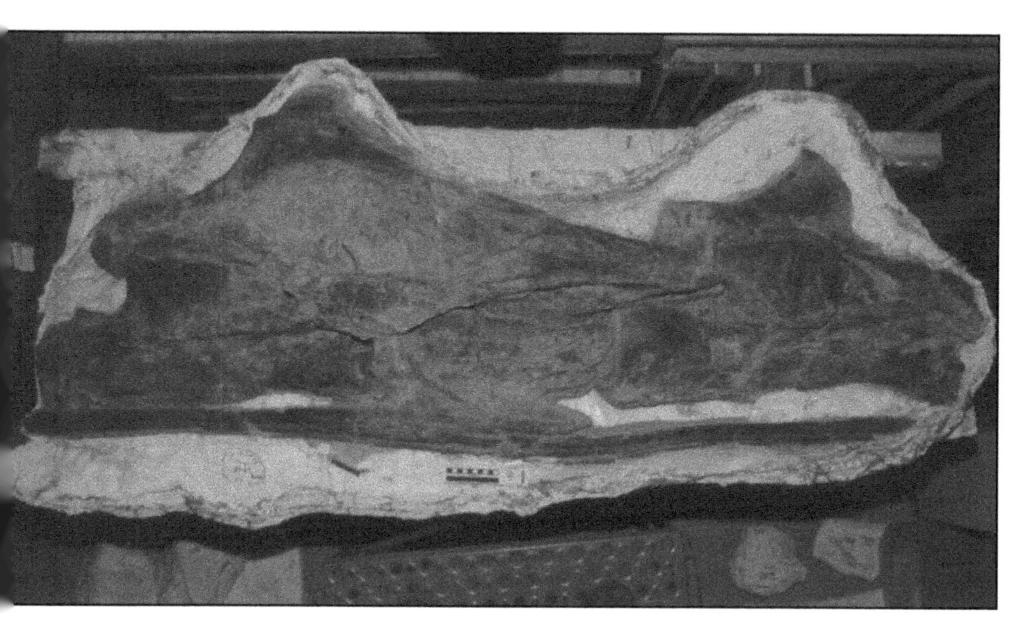

Fig. 129. The enormous cervical vertebrae of *Sauroposeidon proteles*. Photo courtesy of Mathew Wedel.

When Matt and I started corresponding I didn't even know of the existence of MIWG.7306: Matt was contacting me as he was curious as to whether there might be any evidence for giant *Sauroposeidon*-like brachiosaurids in the British Wealden Supergroup. I was therefore very excited when, one day in the summer of 2000, I was able to tell him about my first viewing of MIWG.7306 at David's house. David had been comparing MIWG.7306 with Janensch's figures of *Brachiosaurus*, and while this was still entirely appropriate, I figured that, because *Sauroposeidon* was closer in age to MIWG.7306 than was *Brachiosaurus*, MIWG.7306 might prove to be more like it than was *Brachiosaurus*.

In between other projects (including papers on *Eotyrannus*, *Aristosuchus* and *Thecocoelurus*), I really got to work on the MIWG.7306 paper during 2001, and visited the specimen several more times. While it was obvious which end was which on MIWG.7306, that's about all that was obvious. The thing was a complex mess of bony laminae and concavities (termed fossae), few of which were symmetrical when you compared the two sides. Massive ventrolaterally projecting flanges projected from the bottom surface of the specimen: they were highly similar to the hypertrophied centroparapophyseal laminae regarded by Wedel *et al.* (2000a) as diagnostic for *Sauroposeidon.* My notes and sketches were a complete mess, as every time I looked at the specimen I ended up reinterpreting my previous identifications.

Dinosaurs of the Isle of Wight was finished during May 2001, and a brief comment on MIWG.7306 was added to the sauropod section (Naish & Martill 2001, p. 212). Copies of the book arrived from the Palaeontological Association (the publishers) on 11th July 2001 (the same day that I attended an [unsuccessful] job interview on the Isle of Wight), and at the time I thought that – excepting that brief mention in 'Anon. undated' – this was the first time the specimen had made it into print. How wrong I was – but more on that later.

Completion of the manuscript came slowly, and not until the very end of 2001 was I able to finish the figures. The final manuscript – listing myself, Dave Martill and David Cooper as co-authors – was technically submitted to the chosen journal, *Cretaceous Research,* in February 2002. Why chose *Cretaceous Reseach,* when a journal devoted to vertebrate palaeontology might seem like a more obvious choice? The answer is that the BBC series *Live from Dinosaur Island* was screened in June 2001, and that the production of a special issue of *Cretaceous Research* – devoted entirely to Isle of Wight palaeontology – had been agreed on as a sort of spin-off of the series. A space within that issue was sort of 'booked' for the MIWG.7306 paper.

For various reasons, that special volume never materialised, and the MIWG.7306 manuscript was shelved for what seemed like an eternity, plus the handling editor (who, I should note, was not anyone on the current editorial staff of *Cretaceous Reseach*) managed to lose the original figures of the manuscript. Don't forget that this was in the days before digital submission, so the figures were the original photographs.

So, by February 2002 we had a provisionally complete, submitted manuscript. Mathew Wedel (well known by 2002 as an expert on giant sauropods) and Paul Upchurch (a sauropod expert specialising on phylogenetic relationships) were chosen as reviewers. Their reviews were back by June

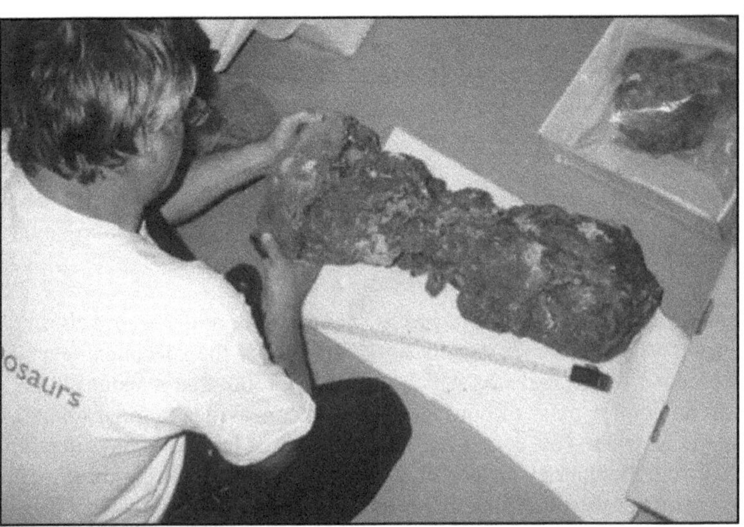

Fig. 130. A second vertebra from the same animal. This one is labeled IWCMS : 2003.28 and is poorly preserved. That's Dave Martill in the photo.

2002, and both were highly positive. Pending minor corrections and additions, this all meant green light to publication.

Looking at it now, I can see that the 2002 version of the manuscript wasn't too bad, but that it could do with improvement and expansion on a few areas. Little known (and perhaps I should have mentioned this earlier) is that a second vertebra was actually discovered at the same place as MIWG.7306, though it's substantially less complete and consists only of a poorly preserved centrum 640 mm long (Fig. 130). This second specimen was mentioned in the pre-review version of the manuscript, but not until November 2003 did I obtain its specimen number (IWCMS : 2003.28). You might be wondering why some specimens at Dinosaur Isle are 'MIWG' while others are 'IWCMS'. Essentially, the former accession code was applied prior to the closure of the old Museum of Isle of Wight Geology, while the latter code was applied to specimens accessioned after the 2001 opening of Dinosaur Isle (IWCMS stands for Isle of Wight County Museum Service).

The figures of the 2002 pre-review version of the manuscript were a bit of a mess. Given that we'd now described the anatomy of MIWG.7306 in detail, a major shortcoming was that the figures didn't point out which bits of the specimen had which names. You really need to do this on sauropod vertebrae, as their anatomy is complex.

In MIWG.7306, three large concavities (termed fossae) occupy most of the lateral surfaces, and these fossae themselves are subdivided into smaller fossae, with some of these subdivisions having further subdivisions. Importantly, the fossae are connected (via openings on their bony walls) to large internal chambers. Bony struts, called laminae, divide the fossae and connect the main 'landmarks' of the bone (such as the pre- and postzygapophyses), and each of these laminae has a name. We presently use a system devised by sauropod expert Jeff Wilson, and his paper on this subject is obligatory reading if you want to understand this nomenclature properly (Wilson 1999).

What are all these fossae and laminae for? Perhaps the most remarkable thing about sauropod vertebrae is how pneumatic they were – the lateral fossae, and the interior of the vertebrae, were occupied, in life, by air sacs. That is, by pneumatic sacs that were connected by tubes both to one another, and to the animal's lungs. How do we know this? Essentially because the detailed anatomy is identical to that of living birds. In birds, bony openings on the sides of the vertebrae also lead into large internal chambers, just as they do in sauropods, and given that the bony openings and internal chambers of bird vertebrae house air sacs, they almost certainly did so in sauropod vertebrae as well. To date no one has proposed a better explanation for the structures seen in sauropods, quite simply because there isn't one. This has obvious implications for mass, as it means that the animals weighed a lot less than you might otherwise think: Wedel (2005) showed that, by calculating for the presence of air sacs, a sauropod comes out at 8-10% lighter than it would have if its mass were uncorrected. This is actually a conservative estimate, as it doesn't account for air sacs that would have been distributed among the soft tissues.

Extensive pneumatisation also has implications for various aspects of sauropod palaebiology

(Wedel 2003). Don Henderson (2003) has worked out what difference it would make for buoyancy were sauropods to go swimming (which they presumably did at times, even though they weren't amphibious or otherwise strongly tied to aquatic environments). Pneumaticity also indicates that sauropods enjoyed the relatively high oxygen extraction levels seen in birds, suggests that sauropods were quite capable of efficient heat dumping, and overall is highly suggestive of an elevated metabolism (as is indicated by the amazingly fast growth style now well documented for sauropods: see Erickson *et al.* 2001, Sander *et al.* 2004). Biologists today seem to think that they're being particularly good, sceptical scientists if they outright reject the silly idea of dinosaurian endothermy, but it's very clear that none of the work that has been done on dinosaur physiology accounts for the amazing degree of pneumatisation seen in sauropods and other groups. There's an awful lot that could be said on the debate over dinosaur physiology, however, and I don't want to tackle it in depth here.

In its degree of pneumaticity, MIWG.7306 appears intermediate between *Brachiosaurus* and *Sauroposeidon*. I worked out that about 60% of each of its lateral surfaces was occupied by the fossae that housed lateral air sacs, and because the specimen is broken in half I could also see that very little of its interior was filled up by internal bony struts: most of it would have been air. The jaw-droppingly amazing thing about the degree of pneumatisation within MIWG.7306 and other brachiosaurids is that their vertebrae consisted of about 80-90% air. Reread that last sentence for emphasis, and you might like to memorise it and trot it out at cocktail parties.

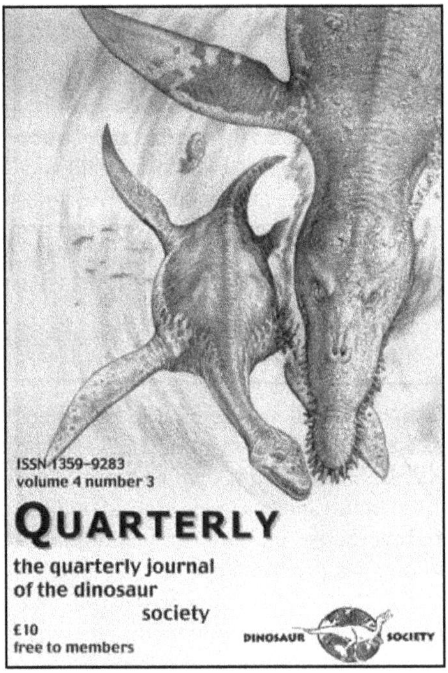

Fig. 131. Cover of the much delayed 2002 issue of *The Quarterley Journal of the Dinosaur Society*. The cover art is by John Sibbick.

Earlier, we saw how the first published measurement of MIWG.7306 gave its length as 920 mm. Measuring a giant sauropod vertebra isn't as simple as you might think: firstly, because the long articular processes (the zygapophyses) may be incomplete, secondly, because the zygapophyses might have become bent or distorted during preservation, and, thirdly, because there's more than one way to measure a vertebra's 'total length'! While MIWG.7306 includes a complete, intact left postzygapophysis, its left prezygapophysis is missing, and its right is broken off. Dave Martill and I re-articulated the broken right prezygapophysis and, measuring from its tip to the posterior-most rim of the posterior articular condyle, came up with a new total length of 1060 mm. That's immense.

In 2002 I was asked to contribute an article on sauropods to an issue of *The Quarterly Journal of the Dinosaur Society*, and I sent it off in late February. For reasons entirely unrelated to the long delay that beset the MIWG.7306 manuscript, this article was also delayed and wasn't published until 2005 (Fig. 131). It included a few paragraphs on MIWG.7306, stated its total length as 1060 mm, and even included a rather fetching photo of me posing with the specimen (Naish 2005). In 2002 I was commissioned to write an article on Isle of Wight dinosaurs for the excellent Japanese magazine *Dino Press* (now, sadly, defunct). Here again I discussed MIWG.7306, again gave its total length as 1060 mm, and again included a silly photo of me (Fig. 132), alongside the specimen (Naish 2002). Amusingly, I've just noticed that the latter article states of MIWG.7306 "It is due to be published some time in 2002, so look out for it!" (English text supplement, p. 25).

Fig. 132. The author poses with MIWG.7306.

Alas, the 1060 mm that I gave in those two articles is, while not technically incorrect, *not* the standardised 'total length' of the specimen for, rather than including prezygapophysis length, the standard way of measuring a sauropod vertebra is to stick to centrum length alone. And the centrum length of MIWG.7306 is 745 mm, which is still highly respectable, and on par with some of the longer cervical vertebrae of HM SII (the larger of the two *Brachiosaurus brancai* specimens described by Janensch). In that specimen, the sixth cervical has a centrum length of 780 mm, while the longest in the series (C10 and C11) each have centrum lengths of 870 mm long. In extra-long *Sauroposeidon*, C8 has a centrum length of 1250 mm while C6 is 1220 mm. Luckily this oversight was caught long before the MIWG.7306 manuscript was resubmitted.

Given that, in June 2002, the MIWG.7306 manuscript had now come back from review, and had fared well during the review process, it was all systems go. But as usual, MIWG.7306 wasn't the only thing I was working on during 2002. We were also working on the *Mirischia* project, the *Leedsichthys* dig at Whittlesey (*Leedsichthys* was a giant Jurassic suspension-feeding fish, equivalent in size to a small whale), the palaeopathology project, and other stuff. So by September of that year, when I attended the 50th SVPCA meeting at the University of Cambridge to gave a talk on MIWG.7306, I still hadn't resubmitted the post-review version of the manuscript.

For the duration of our work on MIWG.7306, Dave had been corresponding with Kent Stevens, best known for his dinomorph software and for his work on the neck posture of sauropods. Dave and Kent had become friends since meeting during the 1998 making of *Walking With Dinosaurs*, and Kent and I had gotten to know each other at SVPCA 1999, held at Edinburgh (SVPCA = Symposium on Vertebrate Palaeontology and Comparative Anatomy).

Kent was of course very interested in MIWG.7306, and when I met up with him again at

SVPCA 2002, one of the things that he, I and Dave Martill discussed was progress with the MIWG.7306 manuscript. To my mild surprise, Dave suggested during conversation that Kent be made a co-author on the manuscript. Don't get me wrong: I was of course very pleased to have Kent in on the project, but I think it was a poor decision of Dave's to bring in another author this late in the day. Doing this would necessitate a major delay, as we'd now have to account for the views and input of another person.

Kent's contribution was useful of course: not only did he produce an excellent, new-look figure of MIWG.7306, he also just happened to have excellent photos of the cervical vertebrae of the *Brachiosaurus brancai* specimen HM SII (originally taken by Chris McGowan, whom you'll know of from Chapter 41). Largely for comparative purposes, we included these figures in the final paper, and they occupy an entire glorious page. In between other projects, Kent and I worked to rehaul the manuscript during 2003 and 2004, and regularly sent lengthy emails to one another. Using various measurements taken off the specimen, we argued extensively over which vertebra it might be (viz, which position it occupied in the neck: brachiosaurids had 13 cervical vertebrae), and what the result might then mean for the total length of the whole animal. The latter sort of calculation is essentially speculative, as we don't know whether proportional neck length was constant across brachiosaurids (Wedel *et al.* 2000). Long-necked *Sauroposeidon*, for example, may not necessarily have had a body that was, proportionally, as big for its neck as was that of *Brachiosaurus*.

Fig. 133. Life restoration of a brachiosaur by Mark Witton.

Anyway, even in June 2004, Kent and I were still talking things through, and we still hadn't declared closure on the manuscript. We finally did just this on June 14th, and on June 23rd the final, post-review, updated, modified version of the article was *finally* submitted. Finally, the paper was in press. You might have noted, incidentally, that I keep fairly reasonable track of dates. I've been keeping diaries since 1997 or so.

Though based only on two vertebrae, only one of which is any good, there is little doubt that MIWG.7306 and IWCMS : 2003.28 represent the largest dinosaur we know of from Britain, and indeed from Europe. It follows then that Dinosaur Isle – the specimen's home repository – hoped to get some nice juicy publicity from the publication of the *Cretaceous Research* paper, and during November 2004 the University of Portsmouth publicity department worked in conjunction with Dinosaur Isle on a planned press release. But here's the problem: *Cretaceous Research* releases its papers when they're still at proof stage, and it releases them as open access on its website. The proofs arrived at the start of October 2004, and I was alerted to the fact that they were available online on 19th November (a Friday). I immediately contacted Dinosaur Isle to let them know, and only at this stage did we produce a press release. The plan was to hold a press conference or something the following week.

But on the morning of Monday 22th I received a phonecall from Paul Rincon, a science reporter for BBC News. He'd seen the on-line *Cretaceous Research* proof and wanted to run a story. I asked if he could wait, given that a press conference and official press release was planned, but, oooh no, that's just not how it works once a story breaks. By 13:40 on that day, the full story, featuring quotes from me and Steve Hutt, was on the BBC News website, and this is where all hell broke loose. The press loved the story, and it was carried in just about every British newspaper, as well as quite a few international ones. I literally spent the entire day on the phone, and I lost track of how many journalists I spoke to. Greg Paul kindly let me use his (now dated) Tendaguru scene depicting a few *Brachiosaurus* individuals, and this was used all over the place in conjunction with photos of MIWG.7306 (though, as always, many publications produced their own god-awful in-house graphics, or used stock images of sauropods, dated c. 1957).

'Britain's biggest dinosaur roamed the Isle of Wight', proclaimed *The Times*; 'Scientists unearth biggest dinosaur to be found in UK', announced *The Scotsman*; and 'Experts bone idle', explained the *Daily Record*. *The Independent*'s article is one of my favourites for its use of hyperbole, as 'Dinosaur bones on Isle of Wight rewrite evolutionary history' seems, even to me, to be just a *little* over-enthusiastic.

While the specimen's size and status proved to be of great interest, also of interest was the fact that the specimen had gone unpublicised for so long. The *Daily Record* title is a direct reference to this, and their article stated "Scientists have [sic] not had enough time to look at the bone, found on the Isle of Wight in 1992. Darren said: "There are thousands of fossils waiting to be studied"". *The Daily Mirror* quoted me as saying "We just hadn't got round to studying it". Even better, after discussing the specimen's remarkable size, the *Daily Telegraph*'s Roger Highfield wrote "Just as remarkable, the neck bone languished in a box for more than a decade before anyone summoned up enough curiosity to study it, according to one of the team, Darren Naish of the University of Portsmouth". Err, somehow I don't think that those were my *exact* words. I also got on local TV, but I sort of lost out by doing the interviews from my house, rather than from Dinosaur Isle, where the specimen is.

Among the many journalists I spoke to at this time, one in particular had something very interesting to say. Unfortunately I forget his name, but I do recall that he was based on the Isle of Wight. His recollection was that the specimen had in fact been brought to the attention of the media before. Basing my conclusions on what happened with *Eotyrannus*, this is almost certainly correct.

Steve Hutt's plan with Isle of Wight dinosaur discoveries (and you'll recall from earlier that Steve was the first person other than Gavin Leng to become acquainted with MIWG.7306) has always been to get publicity both on the discovery of a specimen, *and* on the publication of the formal description. So *Eotyrannus* was in the newspapers as a new dinosaur discovery in 1998, and then again in 2001 when it was formally named and described (Hutt *et al.* 2001). Most of us hold off on talking to the press until our technical work has been published, but I'm not knocking Steve for his double-whammy approach, as the media are evidently interested enough to cover these stories twice.

So while I've never seen the relevant articles, it seems that MIWG.7306 was reported in the newspapers at the time of its discovery. It also turns out that a semi-technical report was published on the specimen, and to my annoyance I didn't find out about this until recently. The article in question is by Jon Radley, well known for his excellent work on Wealden stratigraphy, and it includes two paragraphs on the specimen and a photo (Radley 1997, pp. 108-109). The relevant section reads (though with some typos corrected)...

A brachiosaurid sauropod vertebra from the Wessex Formation (Wealden Group, Lower Cretaceous) of Sudmoor Point

> In the autumn of 1993 Mr. G. Leng discovered a large sauropod vertebra derived from a plant debris bed exposed in the cliff top approximately 1 km northwest of Chilton Chine (SZ 399825). Mr. Leng has generously donated this important specimen to the Museum of IW Geology (MIWG 7306) with the permission of the National Trust. The specimen is preserved in a large, well-cemented sideritic concretion and is consequently only partly crushed. Small quantities of pyrite occur in the bone material and appear to be quite stable.
>
> The bone is 0.75 m long and now ranks as the largest sauropod vertebra in the museum collection. Mr. S. Hutt has identified it as a cervical (neck) vertebra of an adult brachiosaurid sauropod. It is deeply socketed and possesses large prezygapophyses and postzygapophyses. The nature of construction is extremely light with well developed networks of pleurocoels (air chambers). From its size, one can calculate that it came from an animal approximately 22 to 25 m in length. Most brachiosaurid remains discovered so far on the Island are of considerably smaller animals – Radley & Hutt (1993) provided outline details of a recent find. It is hoped that the vertebra will be on temporary display in the museum in the near future.

A few things make Jon's article particularly interesting. The 'autumn of 2003' date he provides is different from the 2002 one I was provided by MIWG staff, and the 750 mm length he provides is of course accurate (and presumably so because it's the centrum length alone). The article also includes the first ever figure of the specimen, showing it in its unprepared, siderite-encased case (a scan of that figure is shown as Fig. 134). Of course the article doesn't in any way diminish the value of the final published description (Naish *et al.* 2004) and is nothing more than an initial, preliminary report. I just wish that I'd known about it when writing *Dinosaurs of the Isle of Wight* and the final MIWG.7306 paper, as then I could have cited it.

Fig. 134. The first ever published figure of MIWG.7306. From Radley (1997).

It turns out that the final published description – while not bad as descriptions go – only really scratches the surface in terms of what information we can learn from MIWG.7306. Previously I discussed the incredible fact that brachiosaurid vertebrae are as much as 80 or 90% air, and that this degree of pneumatisation might mean an awful lot as goes physiology and biology. Because MIWG.7306 is broken into halves, making examination of its pneumatic interior possible, a current project is to get lots more data about pneumaticity out of it. That's ongoing however, so I don't want to talk more about it now.

MIWG.7306 also gives us new information on the diversity and distribution of brachiosaurids. The specimen clearly shares a number of detailed features with *Brachiosaurus* of the Late Jurassic of North America and eastern Africa, and with *Sauroposeidon* of the Early Cretaceous of the USA, and as discussed in the paper (Naish *et al.* 2004) it seems that MIWG.7306 might even be phylogenetically intermediate between these two forms. One of the big questions concerning brachiosaurids is which other sauropods are members of this clade too, and this is an area currently under study by my colleague Mike P. Taylor, who has a lot of new data (and some new species) on this subject.

But if MIWG.7306 represents an animal that is 'phylogenetically intermediate' between *Brachiosaurus* and *Sauroposeidon*, and given that it's been publicised as a 'new' dinosaur, why didn't we name it? That's a good question, and there are two answers.

Firstly, most experts agree that specimens should only be named as new taxa if they can be can be shown to be *diagnostic*: that is, they possess unique features which allow them to be differentiated from other taxa. Given that taxa evolve from ancestors, and evolve into descendants, there aren't sharp boundaries between species and genera – rather, they grade into one another. Consequently, so-called diagnostic features must also, at some stage in any lineage, morph into the slightly different conditions present in ancestors and/or descendants, and multiple intermediate conditions must exist between any two 'diagnostic' end-states. We therefore, arbitrarily, chose cut-off points in how much variation we tolerate within any taxon – in other words, we arbitrarily decide how we chop up a lineage into those units we call species and genera. As a rough rule of thumb this all works, more or less, so long as it's understood that species are artificial segments of lineages (though having said that, any species that can be shown to be the 'end point' of its respective lineage is a clade, and of course this applies to

Fig. 135. The giant brachiosaur represented by MIWG.7306 shown with various contemporaries. *Eotyrannus* (with a dead *Ornithodesmus* in its mouth) stands on a buttress root of a tree while *Aristosuchus* walks by beneath. *Yaverlandia* (here shown as a small pachycephalosaur) stands at middle right and the multituberculate *Loxaulax* lurks at bottom left.

living species given that they don't have descendants). Anyway, I'm going off here at a tangent: this is the sort of heavy philosophical stuff that systematists have been arguing over for decades.

MIWG.7306 possesses loads of anatomical features that are also present in *Brachiosaurus* and *Sauroposeidon*, and it even shares some features that are present in *Sauroposeidon* and not in *Brachiosaurus* but, so far as we can tell, it doesn't possess any features that are unique and thereby allow it to be reliably differentiated from all other sauropods. That's almost certainly an artefact resulting from the fact that we only have two cervical vertebrae of course – if we had the whole skeleton things would be different. But, as it stands, MIWG.7306 cannot presently be diagnosed as a new species.

Actually, for a while I'll admit that we thought that MIWG.7306 *could* be diagnosed. One of several unsolved mysteries about the specimen is the identity of a bizarre, oddly shaped chunk of bone found encased in the same nodule. Smoothly convex on one side, but with a series of subparallel ribs on the other, it was tentatively identified by David Cooper as part of the apex of the neural spine. I thought later on that it might be a partial diapophysis*, as those of brachiosaurids can descend ventrally from the side of the centrum as plate-like processes with smoothly convex lateral surfaces. If this identification is right, then the diapophysis of MIWG.7306 is uniquely odd, as it possesses a small rhomboidal opening near its (presumed) posterior border. Such a feature is unknown elsewhere in sauropods, and it's the sort of feature we might chose to regard as diagnostic. I now have serious doubts about

* In tetrapods with two-headed ribs, the more ventral rib head contacts a facet on the vertebra known as the parapophysis (plural parapophyses), while the more dorsal rib head contacts a facet known as the diapophysis (plural diapophyses). The positions of the parapophyses and diapophyses change along the length of the vertebral column, and indeed their relative positions allow us to identify where in the sequence an isolated vertebra came from.

my identification of this object as a diapophysis, and I think that David was right with the neural spine identification. Images of the object were shown to sauropod-obsessed colleagues, and they remains uncertain as to what it is however, so the thing remains mysterious.

The second answer [to the question: why didn't we name it?] is that MIWG.7306 comes from a geological unit (the Wessex Formation) where there are already lots of named sauropods, virtually all of which are based on non-overlapping fragments, such as vertebrae. What's more, some of these (notably *Eucamerotus*, a form named for dorsal vertebrae), seem to be good honest brachiosaurids closely related to *Brachiosaurus* (and hence to MIWG.7306). It's possible and perhaps likely that some of the other Wessex Formation sauropods, *Eucamerotus* among them, actually represent the same taxon as MIWG.7306, though of course this can't be tested until we have good, more complete specimens (here you'll recall the Barnes High brachiosaurid, still floating in scientific limbo). In view of this situation it would be regarded as bad practise to coin a new name for MIWG.7306.

Partly because it's easier to say than 'MIWG.7306' we've elected to use a totally unofficial nickname for the taxon represented by MIWG.7306, and this is 'Angloposeidon', coined by brachiosaurophile Mike P. Taylor (who is irritated by my continual reference to him as such*).

What do other experts think of MIWG.7306? During the long period of time in which the manuscript was in preparation I spoke to several European colleagues who told me of new sauropods from Portugal and Spain that would easily outclass MIWG.7306 in terms of size. I had this on my mind all the way through the submission process, and at any time I expected there to be some report of a new European sauropod that had a total length exceeding 30 m. But even today such discoveries have yet to materialise, and having now seen some of the specimens in question I know that they fail to come close to the 20 m + estimated for MIWG.7306. 'Angloposeidon' was about equivalent in size to *Brachiosaurus* so far as we know.

The MIWG.7306 paper hasn't been cited much in the literature, but this results from the fact that bugger all has been published on Lower Cretaceous British sauropods since 2004. Matt Wedel – pneumaticity and giant brachiosaurid expert – has certainly been interested and even came to see the specimen in March 2004. He's agreed with our interpretations, and in recent publications on *Sauroposeidon* has noted that MIWG.7306 shows that Britain was once home to a close relative of this Oklahoman giant (Wedel 2005). *Sauroposeidon* is from the Aptian-Albian Antlers Formation, whereas the Wessex Formation that yields MIWG.7306 is just a little older, being late Barremian in age. We also know that similar giant brachiosaurids were present in the Aptian-Albian Cloverly Formation of Montana, as evidenced by a single cervical vertebra held today at the Yale Peabody Museum. The specimen is from a juvenile brachiosaurid but, as Wedel (2005) noted, at 47 cm in length the specimen 'is longer than the vertebrae of many adult sauropods' (p. 55). Footprints produced by *Sauroposeidon*, or by a similar,

---

* So why do I do it? Because there is already a well known Mike Taylor in the world of tetrapod zoology: the marine reptile expert Mike A. Taylor.

closely related brachiosaurid are known from the Glen Rose Limestone of the Paluxy River, Texas. At up a metre in diameter, they must have been produced by a real giant, and, among roughly contemporaneous North American sauropods, only *Sauroposeidon* is big enough (Wedel 2005).

And that's pretty much the whole story up to now. The main message I suppose you should take away is that producing a technical paper on a specimen – even a short one devoted to the description of a single bone – can be an absurdly drawn-out, lengthy affair literally years in the making, and this is all the more so when other projects and life in general get in the way. Discovered by an enthusiastic amateur who donated the specimen to his local museum (Gavin Leng), prepared by an amateur with scientific training (David Cooper), and eventually described technically by a team of palaeontologists, the MIWG.7306 story is also a nice example of the sort of successful collaboration that can result if people work together.

However, there is also a sad ending to this story. David Cooper, who devoted so much time to the specimen and initiated the research that culminated in the paper, had been suffering for some time from cancer, and by 2005 he knew his condition was terminal. Late in June 2005 I received an unexpected and saddening phonecall. Given that I received this news the day before his funeral, I was unable at such short notice to make the arrangements to attend. This, I regret.

Refs - -

- Anon. Undated. *The discovery of the island's largest dinosaur*. Dinosaur Farm Museum (Brighstone, Isle of Wight).
- Erickson, G. M., Curry Rogers, K. & Yerby, S. A. 2001. Dinosaurian growth patterns and rapid avian growth rates. *Nature* 412, 429-433.
- Henderson, D. M. 2003. Tipsy punters: sauropod dinosaur pneumaticity, buoyancy and aquatic habits. *Proceedings of the Royal Society of London B* (Supp.) 271, S180-S183.
- Hutt, S., Naish, D., Martill, D. M., Barker, M. J. & Newbery, P. 2001. A preliminary account of a new tyrannosauroid theropod from the Wessex Formation (Early Cretaceous) of southern England. *Cretaceous Research* 22, 227-242.
- Naish, D. 2002. Thecocoelurians, calamosaurs and Europe's largest sauropod: the latest on the Isle of Wight's dinosaurs. *Dino Press* 7, 85-95.
- Naish, D. 2005. The sauropod dinosaurs of the Wealden succession (Lower Cretaceous) of southern England. *The Quarterly Journal of the Dinosaur Society* 4 (3), 8-11.
- Naish, D. & Martill, D. M. 2001. Saurischian dinosaurs 1: Sauropods. In Martill, D. M. & Naish, D. (eds) *Dinosaurs of the Isle of Wight*. The Palaeontological Association (London), pp. 185-241.
- Naish, D., Martill, D. M., Cooper, D. & Stevens, K. A. 2004. Europe's largest dinosaur? A giant brachiosaurid cervical vertebra from the Wessex Formation (Early Cretaceous) of southern England. *Cretaceous Research* 25, 787-795.
- Radley, J. 1997. Geological report 1993-1994. *Proceedings of the Isle of Wight Natural History and Archaeology Society* 13, 107-114.

- Sander, P. M., Klein, N., Buffetaut, E., Cuny, G., Suteethorn, V. & Le Loeuff, J. 2004. Adaptive radiation in sauropod dinosaurs: bone histology indicates rapid evolution of giant body size through acceleration. *Organisms, Diversity & Evolution* 4, 165-173.
- Taylor, M. P. & Naish, D. 2005. The phylogenetic taxonomy of Diplodocoidea (Dinosauria: Sauropoda). *PaleoBios* 25, 1-7.
- Wedel, M. J. 2003. Vertebral pneumaticity, air sacs, and the physiology of sauropod dinosaurs. *Paleobiology* 29, 243-255.
- Wedel, M. J. 2005. Postcranial skeletal pneumaticity in sauropods and its implications for mass estimates. In Wilson, J. A. & Curry Rogers, C. (eds) *The Sauropods: Evolution and Paleobiology*. University of California Press, Berkeley, pp. 201-228.
- Wedel, M. J. & Cifelli, R.L. 2005. *Sauroposeidon*: Oklahoma's native giant. *Oklahoma Geology Notes* 65, 40-57.
- Wedel, M. J., Cifelli, R. L. & Sanders, R. K. 2000a. *Sauroposeidon proteles*, a new sauropod from the Early Cretaceous of Oklahoma. *Journal of Vertebrate Paleontology* 20, 109-114.
- Wedel, M. J., Cifelli, R. L. & Sanders, R. K. 2000b. Osteology, paleobiology, and relationships of the sauropod dinosaur *Sauroposeidon. Acta Palaeontologica Polonica* 45, 343-388.
- Wilson, J. A. 1999. A nomenclature for vertebral laminae in sauropods and other saurischian dinosaurs. *Journal of Vertebrate Paleontology* 19, 639-653.

Chapter 45:
Finally: big cat kills uncensored and uncut

Something quite important just happened on British TV. Well, important to me. Last week the BBC screened *Big Cat Week*: a series of five programmes broadcast from Kenya's Masai Mari, and featuring the day-to-day lives of a number of individual wild lions, leopards and cheetahs. This is essentially a pared-down version of *Big Cat Diary*, a highly successful TV series the BBC has been showing since September 1996. Screening the individual life histories of particular cats, the series has followed the animals as they have gone about their business hunting, defending territories, courting, mating and raising cubs, and over the years it has become a multi-generational soap-opera as the original stars have died or disappeared, their places being taken by their offspring or by the animals that usurped them. It is easily one of the best things on TV in my opinion – so many remarkable things have been filmed.

As a spin-off for the digital channel BBC Three, the BBC has also been broadcasting *Big Cat Week Uncut*, with each episode being shown immediately after that day's *Big Cat Week* episode. And uncut it truly was. One of things that has always bugged me about nature documentaries is that full acts of predation are almost never shown: you get to see a bit of the chase, then the predators catching their prey, and then a bit of the predators eating the deceased victim. This goes for predation behaviour in lions, hunting dogs, spotted hyaenas, wolves and killer whales.

Why are kills edited? Basically because they are often far more gory than most people expect - kills are often not clean and simple, but may involve the prey animal getting eaten when still alive. Sure, experienced individuals belonging to species that practise precision bites, such as large cats, may well execute a fairly tidy, bloodless killing, but this does not always happen, plus some species – notably hyaenas and hunting dogs – seem to ordinarily take chunks out of the prey while it's still very much alive. In some cases the prey dies from the resulting trauma and blood-loss, and not from tidy bites to its throat or vital organs. Let me add, by the way, that I'm basing these bold assertions on what I've read in books and seen on TV: I have no field experience whatsoever as goes African megafauna. I also want to add that I'm interested in this area, not because I have a love of gore and violence, but because I find macropredation behaviour a fascinating part of the evolutionary biology of the species that practise it.

As you've by now guessed, the big deal about *Uncut* is that it has been showing uncensored kills. To most viewers I hope that these scenes are eye-opening and remarkable, even if gory and horrific, and I have been pleased to see them. Simon King, one of the three presenters*, provides commentary on the footage as he views it. His take on what happens is insightful and accurate, though he has a really annoying habit of describing animals as being 'designed' for certain functions. I met King during 2001 when he was presenting *Live From Dinosaur Island.* We spoke very briefly about co-operative hunting in corvids and raptors, and about *Big Cat Diary*. I got to tell him how awesome I thought the series was, but that was about it.

First introduced to the series when she was a tiny cub in 2005, Duma the cheetah is now about a year old and nearly equal in size to her mother. Both cats were filmed walking through long grass when Duma spotted a Grant's gazelle *Gazella granti* reclining ahead, only its horn tips visible. Alerted by Duma's body language, her mother then began stalking too, and eventually both cats got to within a few metres of the still oblivious gazelle. At up to 80 kg, a male Grant's gazelle is near the upper limit of the cheetah's typical prey choice, though they will also take 90-kg antelopes like topi on occasion, and Hans Kruuk documented cheetahs killing wildebeest and young kudus and zebras.

Almost on top of the gazelle, the cheetahs – very interestingly – deliberately startled it into a run (one cheetah, I forget which, did this by hunching its shoulders and making distinctive stiff-legged bounds in the gazelle's direction). The gazelle didn't have a chance, and both cheetahs were on top of it immediately. After a bit of a struggle, it was pinned down, and Duma began to administer a throat bite. But it seems she wasn't very good at it, and gave up far too early. Experienced cheetahs may require 20 minutes to be confident that they've suffocated their prey, though this might also be the amount of time it takes for them to recover from the exertions of their run (Brakefield 1993). Duma, however, seemed to be holding on for only a few minutes, if that. The gazelle struggled to its feet and had to be pinned down again. Again, Duma tried a throat bite, and again she gave up too soon. But by now her mother had secured a firm purchase on the gazelle's rump, had bitten through the skin, and had started feeding on the animal's haunches. Again the gazelle struggled to its feet, panting and wide-eyed, and again it had to be pinned down. And again another unsuccessful attempt at strangling was performed. As the gazelle stood up again, still struggling and very much alive, it essentially had most of its back end opened up, with a huge bloody hole showing that the mother cheetah had managed to eat in as far as the guts. It was literally being eaten alive. Though the end of the sequence wasn't shown, King explained that the gazelle must have endured a slow, agonising death.

* The other two are Jonathan Scott, well known for his work on the Serengeti, and Saba Douglas-Hamilton: qualified social anthropologist and daughter of elephant expert Iain Douglas-Hamilton.

In *The Velvet Claw* (required watching if you're interested in the evolution of carnivorans: first broadcast by the BBC in 1992), a sequence featuring hostilities between hyaenas and lions showed a similar thing. This time, a lion pride captured a Cape buffalo *Syncerus caffer*. The lions were clearly eating the animal's back end, yet its head was up and it was definitely still alive (if I remember correctly, the animal's head is partially obscured during the sequence by a superimposed 'less alive looking' head!).

During another episode of *Uncut*, King looked in detail at Spotted hyaenas *Crocuta crocuta*, and in particular at the interactions that hyaenas have with lions. As most people know, the two species hate one another, and will kill each other when the opportunity arises. But with one hyaena weighing (at most) 90 kg compared to a female lion's 120-180 kg, and a male's 150-260 kg, hyaenas rarely get to kill adult lions. A lion pride known properly as the Bila Shaka Pride, but dubbed the Marsh Pride for TV, has been a constant presence since filming for *Big Cat Diary* began. Spending most of their time near Musiara Marsh, the pride's history is complex (Scott 2001), as is that of any pride when it's studied for long enough.

During one episode of *Uncut*, a hyaena clan spent a lot of time pestering some of the pride's lionesses. But things went badly wrong for the hyaenas when two male lions showed up. One unfortunate hyaena got stuck in mud: the lions caught it, pinned it down, and one of the lions bit hard into its neck. Smothered in mud, it hung limp in the lion's jaws. But it wasn't dead, and when the lion released it, it clamped onto the lion's upper lip. It was an awesome sequence as, though the hyaena was killed eventually, it endured phenomenal trauma before succumbing. Cats are hard to kill, and I guess hyaenas are too. At the end of the sequence, the lions carried the body out of the mud and water where the fighting had occurred and dumped it on a dry bank. They had no plans to eat the hyaena – they just wanted it dead.

Dean William Buckland once wrote that predatory animals were equipped with sharp teeth and claws so that death was swift and merciful. While nature isn't all red in tooth and claw as some people like to say, predation is often decidedly unpretty, and compassion is a human trait. We might not *like* the idea that predators sometimes start eating their prey while it's still alive, but the fact that it happens is an interesting facet of their behaviour as worthy of knowing as any other.

Refs - -

- Brakefield, T. 1993. *Big Cats: Kingdom of Might.* Voyageur Press (Stillwater, MN).
- Scott, J. 2001. Pride under siege. *BBC Wildlife* 19 (2), 42-48.

Chapter 46:
Meet peccary # 4

Peccaries are predominantly herbivorous, pig-like artiodactyls, restricted today entirely to the Americas, and for reasons that I'll get to in a minute EVERYONE should be talking about them right now. Living species range in weight from 15-40 kg. They are highly social, living in mixed-sex herds of just a few individuals to several hundred, and females produce just one or two precocial babies that follow the mother soon after birth. Peccaries make an interesting assortment of noises: Collared peccaries *Tayassu tajacu* produce loud, dog-like barks, and White-lipped peccaries *T. pecari* scream, bellow and retch when in large groups (small groups tend to be quiet). All species make loud tooth-clacking noises, especially when disturbed.

Peccaries are ecologically flexible, with the three [cough cough] living species being distributed across rainforest, parkland, scrubland, steppe and even desert, and with Collared peccaries in fact occupying all of these habitats. Habits differ according to habitat: rainforest Collared peccaries are diurnal, eat fruit, palm nuts and shrubs, and sleep in burrows, while desert populations are nocturnal, eat mostly cacti, and don't use burrows. This flexibility is reflected in their variable tooth anatomy. Judging from fossils the primitive tooth type for peccaries is bunodonty (viz, where each tooth sports multiple low, rounded, mound-like cusps), but zygodonty (viz, where mound-like cusps are connected by transverse crests) evolved several times. Among living species, the bunodont White-lipped peccary mostly eats nuts while the zygodont Chacoan peccary *Catagonus wagneri* mostly eats cacti. The Collared peccary includes both bunodont and zygodont individuals across its range, and it seems that desert populations are more zygodont while populations from wetter places tend to be bunodont (Wright 1998). Peccaries are reported to occasionally eat carrion, and they will also eat snails and other invertebrates as well as small vertebrates.

The Collared peccary or Javelina *Tayassu tajacu* (or *Pecari tajacu* or *Dicotyles tajacu*) is the best studied species and is the archetypal peccary (Fig. 136), occurring from central Arizona and central Texas south to northern Argentina (though with introduced populations in northern Texas, southern Oklahoma and Cuba). Its nomenclature is a bit confused: some authors use the generic name *Dicotyles* G. Cuvier, 1817 or *Pecari* Reichenbach, 1835 for it, but most common is its inclusion within *Tayassu* Fischer, 1814. *Pecari* is apparently an objective synonym of *Dicotyles* and thus not available, and use of *Dicotyles* therefore depends on whether or not you consider this species distinct enough from the White-lipped peccary to warrant separation. Indeed this confusion is related to a similar controversy over which formal name is used for peccaries: are they Tayassuidae Palmer, 1897 or Dicotylidae Gray, 1868? A few artiodactyl specialists make a point of using the latter name, but the former is more widely used and would easily win in a fight.

Fig. 136. Captive Collared peccaries. These animals were photographed at Marwell Zoo, England, by Mark Witton.

Collared peccaries release an odour like cheese or chicken soup, apparently (Emmons 1997). In fact a vernacular name for them in parts of the USA is musk hog, and you are said to smell them before you see them.

The White-lipped peccary, a species that ranges from southern Mexico to Argentina (and has also been introduced to Cuba), is substantially bigger than the Collared peccary. Mostly an animal of forests, it is semi-nomadic. The third species, the Chacoan peccary, Roman-nosed peccary or Tagua, is particularly notable in being both relatively recently discovered in living state, *and* for being initially named from fossils. I mentioned it before when discussing rodents (see Chapter 21). The species' scientific history began in 1930 when, in his lengthy paper on Argentinian fossil peccaries, C. Rusconi named the new subspecies *Platygonus carlesi wagneri*. By 1948 Rusconi had decided that this form was distinct enough for its own species, *P. wagneri*.

The story then moves on to 1972 when, while working on a mammal inventory project in the semiarid thorn forest and steppe of the Gran Chaco area of Argentina, Paraguay and Bolivia Ralph Wetzel and colleagues were surprised to hear from local people of a large peccary – distinct from the Collared and White-lipped – known to them as the tagua, pagua or curé-bur (meaning donkey-pig). Their enquiries eventually led to the successful procurement of tagua skulls, and they clearly represented a third, modern-day peccary species. Yet again we see a case where good, honest, card-carrying zoologists track down an ethnoknown animal with successful results, or in other words an unarguable example of cryptozoological investigation being carried out by people who don't consider themselves cryptozoologists (for other examples see Chapters 26 and 36). Rather than being new, it now turned out that the tagua was the same thing as Rusconi's fossil species *Platygonus wagneri*: it really was a 'fossil come to life'. But rather than being a member of *Platygonus*, a genus known from the Miocene, Pliocene and Pleistocene of both North and South America, Wetzel concluded that the species was instead better classified within *Catagonus*, a genus first named by Florentino Ameghino in 1904 for Pleistocene Argentinian fossils (Wetzel 1977a, b, Wetzel *et al.* 1975).

The Chacoan peccary is specialised for life in semiarid forests and steppes. It browses on ground cacti, is reported to not drink, and is superior in cursorial ability compared to other living species. Its teeth are particularly tall-crowned and it only has two hind toes, not three like other living peccaries. Reports from hunters suggest that it occurs in several parts of Bolivia where its presence has yet to be verified (Mayer & Wetzel 1986) and it turns out that its fur was being used in the manufacture of New York coats and hats long prior to 1972.

Fig. 137. Captive Chacoan peccary photographed in Phoenix Zoo by Dave Pape.

But here's the big news. While the 1975 publication of live Chacoan peccaries was a major zoological discovery – indeed one of the most significant mammalogical discoveries of the 20th century – it seems that history is repeating itself, for there is now a fourth living peccary species: the Giant peccary. As in the case of the Chacoan peccary, this new species appears to have been discovered by listening to local people: in this case the Caboclos people (descendants of rubber collectors) of the Brazilian Amazon. And the discoverer is Marc van Roosmalen, the Dutch primatologist well known for the many new species of primate he has discovered (about 20) within recent years. After learning of the fabled new peccary, apparently larger than the documented species, van Roosmalen set off with GEO magazine author and film maker Lothar Frenz and two photographers. And after four days of waiting in a hide they were rewarded with views of a group of four of the animals. Good photos were obtained. The animals look distinct from the other living peccaries – they're most like Collared peccaries but larger and without the collar, and they're reported to be even bigger than Chacoan peccaries, hence the name Giant peccary.

German newspapers first reported the successful observation of live Giant peccaries in June 2004, and the news was apparently held back in order to coincide its release with the airing of

Frenz's documentary on the expedition. So far as I can tell, however, Karl Shuker was first to break the news as he mentioned it in his 2002 book *The New Zoo*. Citing personal communication from van Roosmalen, Shuker implied that the Giant peccary had first been encountered in January 2000 (Shuker 2002). It also seems that van Roosmalen and Frenz observed the successful capture and killing of one of the animals: an article in *Suiform Soundings** entitled 'New mammal discovered in South America – and eaten' (Anon. 2004) stated that "Frenz said he and van Roosmalen abstained from trying the meat, but collected some of the remains for a genetic study". The GEO article includes a photo of hunters with a dead Giant peccary, so maybe this is the same individual that Frenz and van Roosmalen watched being eaten.

GEO magazine published an article (in German) on the discovery in 2004, and in December 2005 *Suiform Soundings* published an English version (Carstens 2005). I don't know if van Roosmalen is planning to publish a description based only on the meat sample he collected (species have been described from photos and tissue samples before, the best known case being that of the Bulo Burti boubou *Laniarius liberatus*), or if he's waiting until better material is obtained, but it seems that we have here the valid discovery of a large, terrestrial mammal. That's a big deal, though admittedly not as big a deal as so many people – zoologists included – still seem to think. Multiple new large mammals have been described in recent years, and there's every reason to think that more such discoveries will occur in the future.

Refs - -

- Anon. 2004. New mammal discovered in South America – and eaten. *Suiform Soundings* 4 (2), 66.
- Carstens, P. 2005. Scientist find [sic] new species of large mammal. *Suiform Soundings* 5 (2), 38-39.
- Emmons, L. H. 1999. *Neotropical Rainforest Mammals: A Field Guide (Second Edition)*. University of Chicago Press (Chicago & London).
- Mayer, J. J. & Wetzel, R. M. 1986. *Catagonus wagneri. Mammalian Species* 259, 1-5.
- Shuker, K. P. N. 2002. *The New Zoo*. House of Stratus (Thirsk, North Yorkshire).
- Wetzel, R. M. 1977a. The extinction of peccaries and a new case of survival. *Annals of the New York Academy of Science* 288, 538-544.
- Wetzel, R. M. 1977b. The Chacoan peccary, *Catagonus wagneri* (Rusconi). *Bulletin of the Carnegie Museum of Natural History* 3, 1-36.
- Wetzel, R. M., Dubos, R. E., Martin, R. L. & Myers, P. 1975. *Catagonus*, an 'extinct' peccary alive in Paraguay. *Science* 189, 379-381.
- Wright, D. B. 1998. Tayassuidae. In Janis, C. M., Scott, K. M. & Jacobs, L. L. (eds) *Evolution of Tertiary Mammals of North America. Volume 1: Terrestrial Carnivores, Ungulates, and Ungulatelike Mammals*. Cambridge University Press, pp. 389-401.

* The newsletter of the IUCN/SSC Pigs, Peccaries, and Hippos Specialist Group. Formerly *Asian Wild Pig News*.

Tetrapod Zoology
est. 2006
scienceblogs.com/tetrapodzoology

Index

A

B

E

F

G

H

I

J

K

L

M

N

P

Q

R

S

T

U

Z

ABOUT THE AUTHOR

Dr Darren Naish is Honorary Research Associate at the School of Earth and Environmental Sciences at the University of Portsmouth. He studies dinosaurs and pterosaurs and, in 2001, named the European tyrannosaur *Eotyrannus* (he later described it for his doctoral thesis).

Other notable achievements include his work – written with colleagues – on neck posture in giant, long-necked dinosaurs, and on the life habits of the enormous azhdarchid pterosaurs. He has also worked on giraffes, sea monsters, marine reptiles and fossil seals. He is an avid watcher of living animals at home and abroad and writes as much about living animals as he does about fossil ones. Since 2006, his blog Tetrapod Zoology has become well known as a reliable, entertaining, well-illustrated resource on all matters relating to amphibians, reptiles, birds and mammals, living and fossil. Topics covered have ranged from the global amphibian crisis to giant eagles, poorly known jerboas, horned dinosaurs and mystery big cats. Dr Naish's previous books include *The Great Dinosaur Discoveries*, the *Encyclopedia of Dinosaurs and Prehistoric Life* (co-authored with David Lambert and Elizabeth Wyse), and *Walking With Dinosaurs: The Evidence* (co-authored with David M. Martill).

THE CENTRE FOR FORTEAN ZOOLOGY

So, what is the Centre for Fortean Zoology?

Ve are a non profit-making organisation founded in 1992 with the aim of being a clearing house or information, and coordinating research into mystery animals around the world. We also tudy out of place animals, rare and aberrant animal behaviour, and Zooform Phenomena; ttle-understood "things" that appear to be animals, but which are in fact nothing of the sort, nd not even alive (at least in the way we understand the term).

Why should I join the Centre for Fortean Zoology?

lot only are we the biggest organisation of our type in the world, but - or so we like to think - e are the best. We are certainly the only truly global Cryptozoological research organisation, nd we carry out our investigations using a strictly scientific set of guidelines. We are expand-ig all the time and looking to recruit new members to help us in our research into mysterious nimals and strange creatures across the globe. Why should you join us? Because, if you are enuinely interested in trying to solve the last great mysteries of Mother Nature, there is no-ody better than us with whom to do it.

What do I get if I join the Centre for Fortean Zoology?

or £12 a year, you get a four-issue subscription to our journal *Animals & Men*. Each issue con-ins 60 pages packed with news, articles, letters, research papers, field reports, and even a ossip column! The magazine is A5 in format with a full colour cover. You also have access to ne of the world's largest collections of resource material dealing with cryptozoology and al-ed disciplines, and people from the CFZ membership regularly take part in fieldwork and ex-editions around the world.

How is the Centre for Fortean Zoology organised?

he CFZ is managed by a three-man board of trustees, with a non-profit making trust regis-red with HM Government Stamp Office. The board of trustees is supported by a Permanent irectorate of full and part-time staff, and advised by a Consultancy Board of specialists - many whom are world-renowned experts in their particular field. We have regional representa-ves across the UK, the USA, and many other parts of the world, and are affiliated with other ganisations whose aims and protocols mirror our own.

am new to the subject, and although I am interested I have little practical nowledge. I don't want to feel out of my depth. What should I do?

on't worry. We were *all* beginners once. You'll find that the people at the CFZ are friendly and pproachable. We have a thriving forum on the website which is the hub of an ever-growing ectronic community. You will soon find your feet. Many members of the CFZ Permanent Direc-rate started off as ordinary members, and now work full-time chasing monsters around the orld.

I have an idea for a project which isn't on your website. What do I do?

Write to us, e-mail us, or telephone us. The list of future projects on the website is not exhaustive. If you have a good idea for an investigation, please tell us. We may well be able to help.

How do I go on an expedition?

We are always looking for volunteers to join us. If you see a project that interests you, do not hesitate to get in touch with us. Under certain circumstances we can help provide funding for your trip. If you look on the future projects section of the website, you can see some of the projects that we have pencilled in for the next few years.

In 2003 and 2004 we sent three-man expeditions to Sumatra looking for Orang-Pendek - a semi-legendary bipedal ape. The same three went to Mongolia in 2005. All three members started off merely subscribers to the CFZ magazine.

Next time it could be you!

Project Kerinci, Sumatra - 2003
In search of the bipedal ape Orang Pendek

How is the Centre for Fortean Zoology funded?

We have no magic sources of income. All our funds come from donations, membership fees, works that we do for TV, radio or magazines, and sales of our publications and merchandise. We are always looking for corporate sponsorship, and other sources of revenue. If you have any ideas for fund-raising please let us know. However, unlike other cryptozoological organisations in the past, we do not live in an intellectual ivory tower. We are not afraid to get our hands dirty, and furthermore we are not one of those organisations where the membership have to raise money so that a privileged few can go on expensive foreign trips. Our research teams, both in the UK and abroad, consist of a mixture of experienced and inexperienced personnel. We are truly a community, and work on the premise that the benefits of CFZ membership are open to all.

What do you do with the data you gather from your investigations and expeditions?

Reports of our investigations are published on our website as soon as they are available. Preliminary reports are posted within days of the project finishing.

Each year we publish a 200 page yearbook containing research papers and expedition reports too long to be printed in the journal. We freely circulate our information to anybody who asks for it.

Is the CFZ community purely an electronic one?

No. Each year since 2000 we have held our annual convention - the *Weird Weekend* - in Exeter. It is three days of lectures, workshops, and excursions. But most importantly it is a chance for members of the CFZ to meet each other, and to talk with the members of the permanent directorate in a relaxed and informal setting and preferably with a pint of beer in one hand. Since 2006 - the *Weird Weekend* has been bigger and better and held on the third weekend in August in the idyllic rural location of Woolsery in North Devon.

Since relocating to North Devon in 2005 we have become ever more closely involved with other community organisations, and we hope that this trend will continue. We also work closely with Police Forces across the UK as consultants for animal mutilation cases, and we intend to forge closer links with the coastguard and other community services. We want to work closely with those who regularly travel into the Bristol Channel, so that if the recent trend of exotic animal visitors to our coastal waters continues, we can be out there as soon as possible.

We are building a Visitor's Centre in rural North Devon. This will not be open to the general public, but will provide a museum, a library and an educational resource for our members (currently over 400) across the globe. We are also planning a youth organisation which will involve children and young people in our activities.

Apart from having been the only Fortean Zoological organisation in the world to have consistently published material on all aspects of the subject for over a decade, we have achieved the following concrete results:

- Disproved the myth relating to the headless so-called sea-serpent carcass of Durgan beach in Cornwall 1975
- Disproved the story of the 1988 puma skull of Lustleigh Cleave
- Carried out the only in-depth research ever into the mythos of the Cornish Owlman
- Made the first records of a tropical species of lamprey
- Made the first records of a luminous cave gnat larva in Thailand
- Discovered a possible new species of British mammal - the beech marten
- In 1994-6 carried out the first archival fortean zoological survey of Hong Kong
- In the year 2000, CFZ theories were confirmed when an new species of lizard was added to the British list
- Identified the monster of Martin Mere in Lancashire as a giant wels catfish
- Expanded the known range of Armitage's skink in the Gambia by 80%
- Obtained photographic evidence of the remains of Europe's largest known pike
- Carried out the first ever in-depth study of the *ninki-nanka*
- Carried out the first attempt to breed Puerto Rican cave snails in captivity
- Were the first European explorers to visit the `lost valley` in Sumatra
- Published the first ever evidence for a new tribe of pygmies in Guyana
- Published the first evidence for a new species of caiman in Guyana
- Filmed unknown creatures on a monster-haunted lake in Ireland for the first time
- Had a sighting of orang pendek in Sumatra in 2009
- Published some of the best evidence ever for the almasty in southern Russia

EXPEDITIONS & INVESTIGATIONS TO DATE INCLUDE:

- 1998 Puerto Rico, Florida, Mexico *(Chupacabras)*
- 1999 Nevada *(Bigfoot)*
- 2000 Thailand *(Giant snakes called nagas)*
- 2002 Martin Mere *(Giant catfish)*
- 2002 Cleveland *(Wallaby mutilation)*
- 2003 Bolam Lake *(BHM Reports)*
- 2003 Sumatra *(Orang Pendek)*
- 2003 Texas *(Bigfoot; giant snapping turtles)*
- 2004 Sumatra *(Orang Pendek; cigau, a sabre-toothed cat)*
- 2004 Illinois *(Black panthers; cicada swarm)*
- 2004 Texas *(Mystery blue dog)*
- Loch Morar *(Monster)*
- 2004 Puerto Rico *(Chupacabras; carnivorous cave snails)*
- 2005 Belize *(Affiliate expedition for hairy dwarfs)*
- 2005 Loch Ness *(Monster)*
- 2005 Mongolia *(Allghoi Khorkhoi aka Mongolian death worm)*
- 2006 Gambia *(Gambian sea monster , Ninki Nanka and Armitage's skink*
- 2006 Llangorse Lake *(Giant pike, giant eels)*
- 2006 Windermere *(Giant eels)*
- 2007 Coniston Water *(Giant eels)*
- 2007 Guyana *(Giant anaconda, didi, water tiger)*
- 2008 Russia *(Almasty)*
- 2009 Sumatra *(Orang pendek)*
- 2009 Republic of Ireland *(Lake Monster)*
- 2010 Texas *(Blue dogs)*

THE WORLD'S WEIRDEST PUBLISHING COMPANY

HOW TO START A PUBLISHING EMPIRE

Unlike most mainstream publishers, we have a non-commercial remit, and our mission statement claims that "we publish books because they deserve to be published, not because we think that we can make money out of them". Our motto is the Latin Tag "Pro bona causa facimus" (we do it for good reason), a slogan taken from a children's book `The Case of the Silver Egg` by the late Desmond Skirrow.

WIKIPEDIA: "The first book published was in 1988. `Take this Brother may it Serve you Well` was a guide to Beatles bootlegs by Jonathan Downes. It sold quite well, but was hampered by very poor production values, being photocopied, and held together by a plastic clip binder. In 1988 A5 clip binders were hard to get hold of, so the publishers took A4 binders and cut them in half with a hacksaw. It now reaches surprisingly high prices second hand.

The production quality improved slightly over the years, and after 1999 all the books produced were ringbound with laminated colour covers. In 2004, however, they signed an agreement with LightningSource, and all books are now produced perfect bound, with full colour covers."

Until 2010 all our books, the majority of which are/were on the subject of mystery animals and allied disciplines, were published by `CFZ Press`, the publishing arm of the Centre for Fortean Zoology (CFZ), and we urged our readers and followers to draw a discreet veil over the books that we published that were completely off topic to the CFZ.

However, in 2010 we decided that enough was enough and launched a second imprint, `Fortean Words` which aims to cover a wide range of non animal-related esoteric subjects. Other imprints will be launched as and when we feel like it, however the basic ethos of the company remains the same: Our job is to publish books and magazines that we feel are worth publishing, whether or not they are going to sell. Money is, after all - as my dear old Mama once told me - a rather vulgar subject, and she would be rolling in her grave if she thought that her eldest son was somehow in `trade`.

Luckily, so far our tastes have turned out not to be that rarified after all, and we have sold far more books than anyone ever thought that we would, so there is a moral in there somewhere...

Jon Downes,
Woolsery, North Devon
July 2010

Other Books in Print

The Mystery Animals of Ireland by Gary Cunningham and Ronan Coghlan
Monsters of Texas by Gerhard, Ken
The Great Yokai Encyclopaedia by Freeman, Richard
NEW HORIZONS: Animals & Men *issues 16-20 Collected Editions Vol. 4* by Downes, Jonathan
A Daintree Diary - Tales from Travels to the Daintree Rainforest in tropical north Queensland, Australia by Portman, Carl
Strangely Strange but Oddly Normal by Roberts, Andy
Centre for Fortean Zoology Yearbook 2010 by Downes, Jonathan
Predator Deathmatch by Molloy, Nick
Star Steeds and other Dreams by Shuker, Karl
CHINA: A Yellow Peril? by Muirhead, Richard
Mystery Animals of the British Isles: The Western Isles by Vaudrey, Glen
Giant Snakes - Unravelling the coils of mystery by Newton, Michael
Mystery Animals of the British Isles: Kent by Arnold, Neil
Centre for Fortean Zoology Yearbook 2009 by Downes, Jonathan
CFZ EXPEDITION REPORT: Russia 2008 by Richard Freeman *et al*, Shuker, Karl (fwd)
Dinosaurs and other Prehistoric Animals on Stamps - A Worldwide catalogue by Shuker, Karl P. N
Dr Shuker's Casebook by Shuker, Karl P.N
The Island of Paradise - chupacabra UFO crash retrievals, and accelerated evolution on the island of Puerto Rico by Downes, Jonathan
The Mystery Animals of the British Isles: Northumberland and Tyneside by Hallowell, Michael J
Centre for Fortean Zoology Yearbook 1997 by Downes, Jonathan (Ed)
Centre for Fortean Zoology Yearbook 2002 by Downes, Jonathan (Ed)
Centre for Fortean Zoology Yearbook 2000/1 by Downes, Jonathan (Ed)
Centre for Fortean Zoology Yearbook 1998 by Downes, Jonathan (Ed)
Centre for Fortean Zoology Yearbook 2003 by Downes, Jonathan (Ed)
In the wake of Bernard Heuvelmans by Woodley, Michael A

CFZ EXPEDITION REPORT: Guyana 2007 by Richard Freeman *et al*, Shuker, Karl (fwd)
Centre for Fortean Zoology Yearbook 1999 by Downes, Jonathan (Ed)
Big Cats in Britain Yearbook 2008 by Fraser, Mark (Ed)
Centre for Fortean Zoology Yearbook 1996 by Downes, Jonathan (Ed)
THE CALL OF THE WILD - Animals & Men issues 11-15
Collected Editions Vol. 3 by Downes, Jonathan (ed)
Ethna's Journal by Downes, C N
Centre for Fortean Zoology Yearbook 2008 by Downes, J (Ed)
DARK DORSET -Calendar Custome by Newland, Robert J
Extraordinary Animals Revisited by Shuker, Karl
MAN-MONKEY - In Search of the British Bigfoot by Redfern, Nick
Dark Dorset Tales of Mystery, Wonder and Terror by Newland, Robert J and Mark North
Big Cats Loose in Britain by Matthews, Marcus
MONSTER! - The A-Z of Zooform Phenomena by Arnold, Neil
The Centre for Fortean Zoology 2004 Yearbook by Downes, Jonathan (Ed)
The Centre for Fortean Zoology 2007 Yearbook by Downes, Jonathan (Ed)
CAT FLAPS! Northern Mystery Cats by Roberts, Andy
Big Cats in Britain Yearbook 2007 by Fraser, Mark (Ed)
BIG BIRD! - Modern sightings of Flying Monsters by Gerhard, Ken
THE NUMBER OF THE BEAST - Animals & Men issues 6-10
Collected Editions Vol. 1 by Downes, Jonathan (Ed)
IN THE BEGINNING - Animals & Men *issues 1-5 Collected Editions Vol. 1* by Downes, Jonathan
STRENGTH THROUGH KOI - They saved Hitler's Koi and other stories by Downes, Jonathan
The Smaller Mystery Carnivores of the Westcountry by Downes, Jonathan
CFZ EXPEDITION REPORT: Gambia 2006 by Richard Freeman *et al*, Shuker, Karl (fwd)
The Owlman and Others by Jonathan Downes
The Blackdown Mystery by Downes, Jonathan
Big Cats in Britain Yearbook 2006 by Fraser, Mark (Ed)
Fragrant Harbours - Distant Rivers by Downes, John T
Only Fools and Goatsuckers by Downes, Jonathan
Monster of the Mere by Jonathan Downes
Dragons:More than a Myth by Freeman, Richard Alan
Granfer's Bible Stories by Downes, John Tweddell
Monster Hunter by Downes, Jonathan

The Centre for Fortean Zoology has for several years led the field in Fortean publishing. CFZ Press is the only publishing company specialising in books on monsters and mystery animals. CFZ Press has published more books on this subject than any other company in history and has attracted such well known authors as Andy Roberts, Nick Redfern, Michael Newton, Dr Karl Shuker, Neil Arnold, Dr Darren Naish, Jon Downes, Ken Gerhard and Richard Freeman.

Now CFZ Press is launching a new imprint. Fortean Words is a new line of books dealing with Fortean subjects other than cryptozoology, which is - after all - the subject the CFZ are best known for. Fortean Words is being launched with a spectacular multi-volume series called *Haunted Skies* which covers British UFO sightings between 1940 and 2010. Former policeman John Hanson and his long-suffering partner Dawn Holloway have compiled a peerless library of sighting reports, many that have been made public before.

Other forthcoming books include a look at the Berwyn Mountains UFO case by renowned Fortean Andy Roberts and a series of books by transatlantic research Nick Redfern.

CFZ Press is dedicated to maintaining the fine quality of their works with Fortean Words. New authors tackling new subjects will always be encouraged, and we hope that our books will continue to be as ground breaking and popular as ever.